ALGEBRA FROM A TO Z

ALGEBRA FROM A TO Z

Volume 2

A. W. Goodman
University of South Florida, USA

Published by

World Scientific Publishing Co. Pte. Ltd.
P O Box 128, Farrer Road, Singapore 912805
USA office: Suite 1B, 1060 Main Street, River Edge, NJ 07661
UK office: 57 Shelton Street, Covent Garden, London WC2H 9HE

British Library Cataloguing-in-Publication Data
A catalogue record for this book is available from the British Library.

ALGEBRA FROM A TO Z (Vol. 2)

ISBN 981-02-4478-9 (Set)
ISBN 981-02-4980-2 (Vol. 2)

Printed in Singapore by Mainland Press

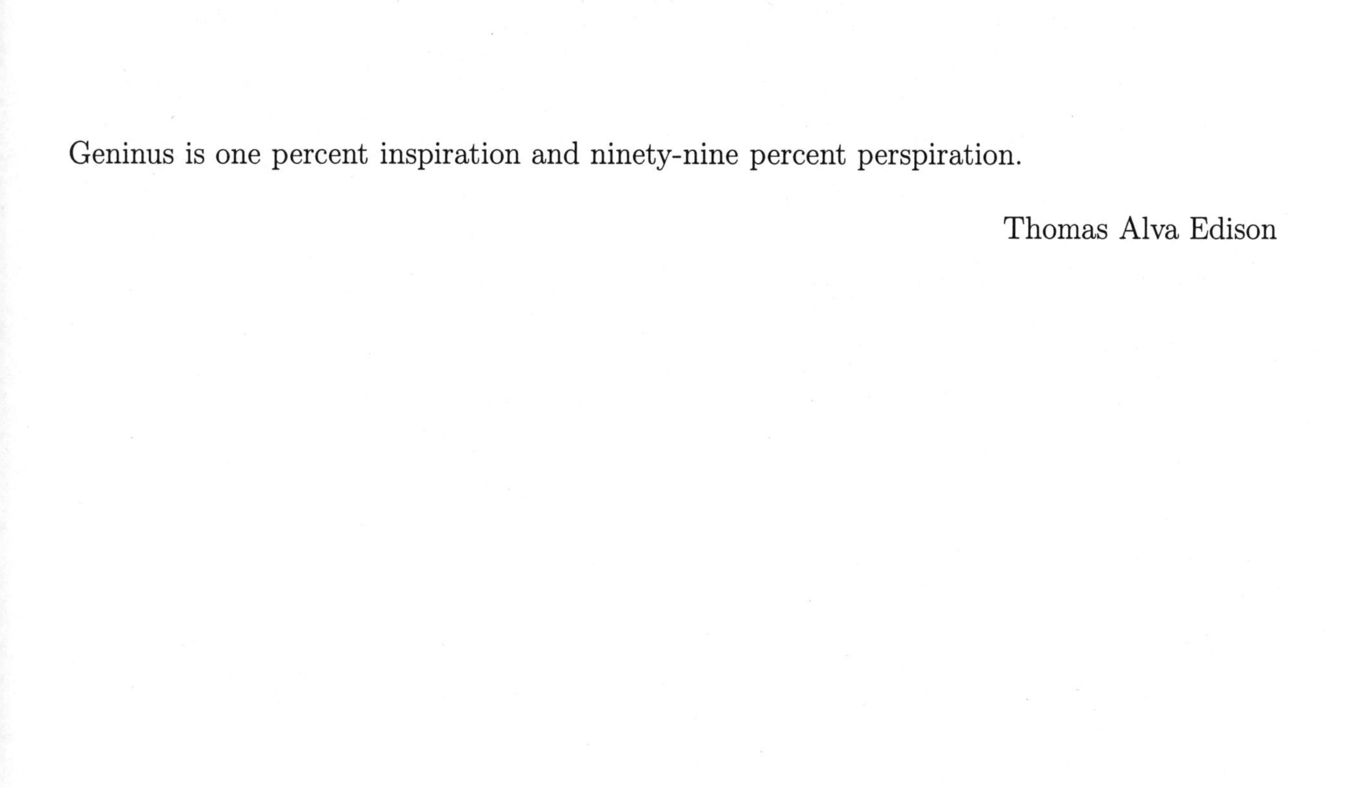

Geninus is one percent inspiration and ninety-nine percent perspiration.

Thomas Alva Edison

Preface

There are many books on Algebra so why must we have another one? These booklets are *not* a beginning algebra book, although *they start from the very beginning.* They are aimed at the student who has had some algebra, or is taking the course now, but does not really understand what is going on. We hope that by reading these booklets the confused student will find most of his/her difficulties removed and many vague ideas clarified.

Most books *explain very little.* The authors are content to hit the high spots, leaving many details for the teacher to explain in person. But the real difficulty lies deeper. Pure mathematics consists of a sequence of definitions, theorems and proofs. This formal sequence often leaves the reader very confused. In the face of such formality, the student should be asking "What is this?", or "Why do I want to learn this?" or, even worse "Who cares about this?". The other extreme is a presentation of the down-to-earth explanations (called motivation) which is not really mathematics. A good book should contain both the motivation (down-to-earth explanations) and the pure mathematical material which is the ultimate goal. To provide both requires more space and time than is usually available. The result is that every algebra book that I have seen gives neither (a) the motivation nor (b) the pure mathematical results (definitions, theorems, and proofs). Instead these algebra texts present a poor mixture of the two. Perhaps during a first view of algebra, such a mixture is unavoidable, but to truly understand algebra, both (a) and (b) should receive full recognition.

Further, most textbook authors are afraid to give the real down-to-earth explanations for fear that their colleagues will criticize the author for lack of rigor. But the same authors are often afraid to give the pure mathematical explanation for fear of frightening the student and losing him/her. In this book we are not afraid to tell the student what is really going on by giving the down-to-earth presentation. Then we follow this with the correct mathematical treatment, which would probably leave the student puzzled if he did not see the motivation first. The progress from the real "down-to-earth explanation" to the "pure mathematical presentation" is sometimes a rocky road that must be traveled by the serious student — but it should be traveled slowly.

This type of two-fold presentation takes more space than is permitted in the usual algebra text. Our solution to this problem is to divide the material into five little booklets. Thus, the reader can pick and choose the particular topics of interest, and buy only the one or two that suits his/her needs. In this way the student will not only save money, but will also save the energy required to lug around a 1500 page monster that includes material that he/she does not want and will not read.

Can these booklets be used as a textbook for a course in Elementary Algebra either in a high school, or in college, or in a University? The answer is YES.

If these booklets are used as a text some care must be observed. The first three chapters are too simple for classroom presentation. They are included because in *Algebra from A to Z*, we must start at the beginning. The teacher should assign the first three chapters as outside reading, meaning that the student reads and studies the first three chapters on his/her own. A few days can be devoted to answering questions that the student might have. Or the teacher might ask questions of the students to be certain that they have read and understood the first three chapters. The course then starts on Chapter 4. The teacher can go as fast or as slow as the local conditions and previous education demand. Of course the teacher can make whatever deletions or additions he or she may feel are appropriate, but I believe that all of the important items in elementary algebra are included in the remaining chapters. In fact, there are a few special topics that are not usually found in an elementary algebra text. The outstanding example is the last chapter which contains a thorough discussion of the Peano Axioms. This material is not available in any elementary textbook.

This is the proper place to acknowledge the help I have received. First I must thank Dr. Stanley Liu and Dr. Jitan Lu. Without their kind encouragement and constant attention to the production of these booklets, they would never have appeared in print. I also owe a great debt to William Goodman who helped with the proofreading and uncovered many errors. Finally I must mention that Ms. Janine Mattson read the entire manuscript and made many important suggestions for improving the presentation of the material.

A. W. Goodman
Tampa, Florida

Contents

Chapter 6

Functions and Linear Equations

1. Functions and Function Notation

This chapter is devoted to the study of linear equations and their applications. However it will help our ability to state our results precisely if we can use function notation. The concept of a function and the use of function notation will be studied thoroughly in a later chapter. Here we cover only the basic ideas that we will need in this chapter.

The reader has already met functions quite frequently, but he was probably not aware of this because of foggy conditions at the time of the encounter. The square of a number is a function of the number. The number of distinct prime factors of a positive integer is a function of that integer. Any formula such as

$$y = x^2 - x^3 \tag{1}$$

gives a function. The following definition may look complicated and terrifying, but is really very simple.

Definition 1 (Function). Let A and B be any two non-empty sets. If for each x in A there is a rule (method or procedure) that gives a unique y in B, then the rule (method or procedure) determines a *function* from the set A to the set B.

Of course we can use any letter to represent an element of A but x is the standard one and x (or whatever letter is used) is called the *independent variable.* Similarly y is the standard letter for the elements of B, and (whatever letter is used) is called the *dependent variable.*

One more bit of new notation is necessary before we give some concrete examples. We need a notation to indicate the relation that y is a function of x. The standard collection of symbols (used all over the civilized world) is

$$y = f(x)\,, \tag{2}$$

which is read "$y = f$ of x". Of course different letters can be used for different functions. For example popular letters for functions are f, g, h, F, G, H, and the Greek letters ϕ, φ, and ψ. Please note that in Equation (2) the f does NOT multiply the x (as our previous use might indicate). If we need a mental picture of Equation (2) we can think of f as a machine, or an operator that works on x and produces a corresponding y. This machine or operator will be different for different functions.

We have had a rather heavy dose of theory. Now we look at some examples and for this we return to the functions mentioned at the beginning of this section.

If $f(x)$ is the squaring function we may write

$$y = f(x) = x^2 . \tag{3}$$

Suppose our function is the number of distinct prime factors of x. We have no formula for computing this number but we can always introduce a new symbol such as

$$y = n(x) , \tag{4}$$

which is read "$y = n$ of x".

If our function is the one defined by the formula in Equation (1) then (using a different letter for a new function) we may write

$$y = g(x) = x^2 - x^3 . \tag{5}$$

How does this new notation work in a practical situation? This is

Example 1. If $f(x) = x^2$, find the function (the corresponding y) when $x = 2$, 5, 0, 1/3, -4, and π.

Solution. We merely replace x by the number in question and compute y (if possible). Following these instructions we have for y:

$$f(2) = 2^2 = 4 , \qquad f(5) = 5^2 = 25 , \qquad f(0) = 0^2 = 0 ,$$
$$f(1/3) = (1/3)^2 = 1/9 , \quad f(-4) = (-4)^2 = 16 , \quad f(\pi) = \pi^2 . \qquad ▶▶$$

Notice that when $x = \pi$, we cannot compute π^2 by giving ALL the digits in its decimal representation. If an approximation is acceptable, then a computation is possible. Otherwise $y = \pi^2$ is the best we can do.

Example 2. If the function is $n(x)$, the number of distinct prime factors of x, find $y = n(x)$ when $x = 6$, 12, 30, 121, 2048, and 2310.

Solution. In this problem we must first agree on the nature of the number 1. In mathematics the number 1 is not considered to be a prime factor of any number. With this agreement we have

$$\begin{aligned}
n(6) &= 2 , && \text{because } 6 = (2)(3) , \\
n(12) &= 2, && \text{because } 12 = (2)^2(3) \text{ and the factor 2 is only counted once} , \\
n(30) &= 3 , && \text{because } 30 = (2)(3)(5) , \\
n(121) &= 1 , && \text{because } 121 = (11)^2, \\
n(2048) &= 1 , && \text{because } 2048 = (2)^{11}, \\
n(2310) &= 5 , && \text{because } 2310 = (2)(3)(5)(7)(11).
\end{aligned}$$
▶▶

Example 3. If $y = g(x) = x^2 - x^3$, the function given by Equation (5), find y when $x = 1$, 2, 3, 5, 1/2, -1, and -5.

Solution. According to the meaning of the symbols we replace x by the number given and compute. Arithmetic gives:

$$y = g(1) = 1 - 1 = 0\,,$$
$$y = g(2) = 2^2 - 2^3 = 4 - 8 = -4\,,$$
$$y = g(3) = 3^2 - 3^3 = 9 - 27 = -18\,,$$
$$y = g(5) = 5^2 - 5^3 = 25 - 125 = -100\,,$$
$$y = g(1/2) = (1/2)^2 - (1/2)^3 = 1/4 - 1/8 = 1/8\,,$$
$$y = g(-1) = (-1)^2 - (-1)^3 = 1 + 1 = 2\,,$$
$$y = g(-5) = (-5)^2 - (-5)^3 = 25 + 125 = 150\,.$$ ►►

Example 4. If $f(x) = x^2$ find $f(a+b)$ and $f(x+y)$.

Solution. The symbols $f(a+b)$ resemble multiplication and we are tempted to use the distributive law and write $f(a+b) = f(a)+f(b) = a^2+b^2$. But this is *WRONG*. To find the correct answer we replace x by $a+b$ and compute, obtaining $f(a+b) = (a+b)^2 = a^2+2ab+b^2$, and this is the correct answer.

Suppose that for the function of Example 4 we want $f(x+y)$ and we write $y = (x+y)^2$ so that $y = x^2 + 2xy + y^2$. *THIS IS COMPLETELY WRONG*, but the casual reader may have trouble spotting the error. We have used the symbol y with two different meanings in the same problem. First we used it to represent the dependent variable, and then we used it again in $x+y$ as a part of the independent variable. The correct form for the answer is: if $f(x) = x^2$ then $f(x+y) = x^2 + 2xy + y^2$, or simply $(x+y)^2$. ►►

SLOGAN. Never use one symbol with two different meanings in the same problem.

Now for a little more theory and two new words. When a function is defined on a set A, the set A is called the *DOMAIN* of the function and is denoted by D (not necessarily the set of real numbers). As the independent variable x runs through all elements of the set A, the values of y need not fill out the image set B. The set that it does generate is called the *RANGE* of the function and is denoted by R (not always the set of real numbers).

When a function is given, the domain of the function should also be specified. This necessary detail is frequently omitted to save time and space. This is not too serious because of the following

AGREEMENT. If the domain of a function is not specified, then the domain is the largest set for which the definition makes sense.

Example 5. Find the domain and range for the functions given in Examples 1, 2, and 3.

Solution. The function $f(x) = x^2$ is defined for all real numbers and therefore the domain is the set of all real numbers. But x^2 is never negative so the range is the set of all $y \geq 0$.

The function $n(x)$ is only defined for positive integers. Hence this is the domain. But for any given positive integer N, there is an integer x which has N distinct prime factors. If N

is large then x will be very large, but this has no effect on the problem at hand. The range of the function $n(x)$ is the set of all positive integers.

The domain of $f(x) = x^2 - x^3$ is the set of all real numbers. If we look at the few values we computed it is easy to guess that the range of this $f(x)$ is also the set of all real numbers. ▶▶

Notice that we did not PROVE this last statement, nor were we asked to prove it. We were asked only to guess (correctly) at the range. A proof requires a new concept, namely continuity, and this item belongs in calculus.

Exercise 1

In Problems 1 through 10 a function is defined by the given formula. In each case find $f(x)$ for the values of x listed.

1. $f(x) = 2x + 3, \quad x = -2, -1, 0, 1, 2, 10.$

2. $g(x) = 3x - 2, \quad x$ as in Problem 1.

3. $h(x) = x^2 + 5, \quad x$ as in Problem 1.

4. $F(x) = \dfrac{10}{x^2 + 5}, \quad x$ as in Problem 1.

5. $G(x) = \dfrac{10}{3x - 2}, \quad x$ as in Problem 1.

6. $H(x) = x^2 - 3x + 2, \quad x$ as in Problem 1.

Hint: Should you factor $f(x)$ before you compute?

$$H(x) = (x - 1)(x - 2).$$

7. $f(x) = x^2 - 6x + 8, \quad x = 0, 1, 2, 3, 4, 5, 6.$

8. $f(x) = x^2 + 7x + 12, \quad x = -7, -6, -5, -4, -3, -2, -1.$

9. $f(x) = x(x - 1)(x - 2), \quad x = -2, -1, 0, 1, 2, 3, 4.$

10. $f(x) = x(x + 1)(x + 2)(x + 3), \quad x = -3, -2, -1, 0, 1.$

In Problems 11 through 15 a function is described in words. Find a formula for the given function.

11. The surface area S of a cube is a function of the length x of one of its edges.

12. The volume V of a cube is a function of the length x of one of its edges.

13. The volume of a cube is a function of S the surface area.

14. The area A of an equilateral triangle is a function of x, the length of one of its sides.

15. The area of an equilateral triangle is function of H, the length of one of its altitudes.

***16.** Let $d(x)$ be the number of divisors of the integer $x > 0$. Find $d(x)$ for $x = 3$, 4, 12, 19, 30, and 60. In this problem both 1 and x are divisors of x.

***17.** Let $s(x)$ be the sum of the divisors of the positive integer x. Here 1 is considered to be a divisor of any integer. Find $s(x)$ for $x = 3$, 4, 6, 12, 19, and 30.

***18.** Let $\pi(x)$ be the number of primes less than or equal to x. For this Problem 1 is not considered to be a prime. Thus $\pi(2) = 1$. Find $\pi(6)$, $\pi(9)$, $\pi(25)$, $\pi(\sqrt{98})$ and $\pi(\sqrt{120})$.

In Problems 19 through 28 find the domain and range of the function that is defined in the earlier problems in this set.

19. Problem 1.

20. Problem 2.

21. Problem 3.

***22.** Problem 4.

***23.** Problem 5.

***24.** Problem 6.

25. Problem 7.

26. Problem 11.

27. Problem 12.

28. Problem 18.

***29.** Find a function $f(x)$ that is zero for $x = -4, -3, -2, -1, 0, 1, 2, 3, 4$, but $f(5) \neq 0$. What is $f(5)$ for your function?

30. If $g(x) = x^2 + 3x - 7$, find $g(1)$, $g(4)$, and $g(5)$. Is it true $g(1) + g(4) = g(5)$?

31. For $g(x)$ from Problem 30, find $g(c + 2)$ and simplify the result.

32. If $f(x) = (5 + x)/(11 - x)$, find $f(x + 4)$.

33. If $F(y) = (6 + y)/(8 - y)$, find $F(z + 3)$.

34. If $H(a) = (5 + a)/(11 - a)$, find $H(a + 4)$.

35. If $J(y) = y^2 - 3y$, find $J(y^2 - 2)$ and simplify the result as much as possible.

36. If $f(x) = 4x^2 - 7$ find $f(3x) - 2f(x + 3)$ and simplify the result as much as possible.

37. If $g(x) = 3x^2 + 11$, find $5g(x) - 2g(x + 5)$ and simplify result as much as possible.

38. If $f(x) = 6x^2 + 5$, find $f(2x) - 4f(x + 2)$ and simplify.

---

2. Equations

Before we look at linear equations, we will make a few remarks about equations in general. If $f(x)$ and $g(x)$ are two functions then the statement

$$\boxed{f(x) = g(x)} \tag{6}$$

is an equation. If the statement is true for all x in the common domain of the two functions then the statement (6) is called an *identity.* If this is not the case, then (6) is called a *conditional equation.* We have already discussed these two concepts in Chapter 1, but a little review is in order. For example, the equation

$$x^3 - 9x = x(x-3)(x+3) \tag{7}$$

is an identity because it is true for all real x, (try a few yourself).

The equation

$$x^3 - 9x = x^2 - 9x \tag{8}$$

is a conditional equation because it is true if $x = 0$ and if $x = 1$, and for no other values of x. In this case the two numbers are called *solutions* of Equation (8). A set is often denoted by placing the members inside braces. In this case we write $\{0, 1\}$ for the solutions of (8). The set of all solutions of an equation is called *the solution set.*

Several hundred years ago the main problem of algebra was "how to solve equations". Very efficient methods were developed, and "modern" algebra is now concerned with very different problems. However solving equations is still an important technique in every branch of science. Here we will look at a few of the more elementary operations that we will be using to solve equations.

Suppose that when $x = r$ the relation

$$f(r) = g(r) \tag{9}$$

is true. Then r is called a *root* of Equation (6). If $h(x)$ is any function and we add $h(x)$ to both sides of Equation (9) we will have

$$f(r) + h(r) = g(r) + h(r) \tag{10}$$

so that r is also a root of

$$f(x) + h(x) = g(x) + h(x)\,. \tag{11}$$

Conversely, if r satisfies Equation (10) and we subtract $h(r)$ from both sides, then we obtain Equation (9) and r is also a root of (6). This gives us

Theorem 1. *Let $h(x)$ be any function. The number r is a root of*

$$f(x) = g(x)\,, \tag{6}$$

if and only if r is also a root of

$$f(x) + h(x) = g(x) + h(x)\,. \tag{11}$$

In other words, the solution set of Equation (6) is the same as the solution set of Equation (11). When this occurs for two different equations, the two equations are said to be *equivalent.* Thus we have

Rule 1 (Slogan). In trying to solve an equation we can add the same function to both sides of the equation without harm, since any solution of the first equation is also a solution of the second equation and conversely.

Now we give a few comments.

First, since $h(x)$ may be negative, Rule 1 also includes subtraction.

Second, the same rule does *NOT* apply for multiplication or division. Some modification is necessary if a rule for products or quotients is to be correct. More about this later in this section.

Third, notice that in Theorem 1 or in the Rule 1 nothing is said about the domains of the functions $f(x)$, $g(x)$, and $h(x)$. It is obvious that the domains of all three functions must include the roots we are seeking. Otherwise the statements are pure nonsense. To include some statement about the domains in the theorem or rule, would only make the statements more complicated and harder to understand, and our goal is understanding mathematics. In the future we will always omit any unnecessary items such as "all of the functions considered must have the same domain".

Theorem 2. *If $F(x)$ is the product of two functions,*

$$\boxed{F(x) = A(x)B(x)\,,} \tag{12}$$

then the equation

$$\boxed{F(x) = 0} \tag{13}$$

has a solution $x = r$, if and only if either $A(r) = 0$ or $B(r) = 0$, or both $A(r) = 0$ and $B(r) = 0$.

Proof. It is sufficient to observe that in the real number system the product ab of two numbers is 0, if and only if either $a = 0$, or $b = 0$, or both a and $b = 0$. ∎

We have had a heavy dose of theory. Let us see how EASY it is to apply this theory. To solve equation (8): $x^3 - 9x = x^2 - 9x$, we add the same $f(x) = -x^2 + 9x$ to both sides and use Theorem 1. This addition will give

$$x^3 - 9x + (-x^2 + 9x) = x^2 - 9x + (-x^2 + 9x)\,. \tag{14}$$

Certain terms add to 0, for example on the right side $-9x + 9x = 0$. We also call this cancellation (or cancellation by addition). We are left with the equivalent equation

$$x^3 - x^2 = 0\,. \tag{15}$$

If we factor the left side of (15) and use Theorem 2, we will have $x^3 - x^2 = x^2(x - 1) = 0$ so either $x = 0$ or $x = 1$. The solution set for equation (8) is $\{0, 1\}$.

In solving equation (8) it is not necessary for you to write Equation (14), because this simple step can be done in your head. It is sufficient to move the function on the right side of (8) to the left side and supply a minus sign. Sometimes this operation is called "*transposing*", so we transpose $x^2 - 9x$ and change the sign of each one of the terms. This will give Equation (15) more quickly.

Example 1. Solve the equation (find the solution set for)

$$3x^2 + 7x - 40 = 2x^2 + 5x - 25\,. \tag{16}$$

Solution. Here we subtract the function on the right side from both sides (or transpose and supply a minus sign). This gives

$$3x^2 + 7x - 40 - (2x^2 + 5x - 25) = 0\,,$$
$$x^2 + 2x - 15 = 0\,,$$

and on factoring (see Chapter 5, Section 1) we have

$$(x+5)(x-3) = 0\,.$$

Either $x + 5 = 0$, $x = -5$, or $x - 3 = 0$, $x = 3$.

The solution set for Equation (16) is $\{-5, 3\}$. ▶▶

More theory. Suppose that instead of adding something, we *multiply* both sides of

$$f(x) = g(x) \tag{6}$$

by $h(x)$, to obtain

$$h(x)f(x) = h(x)g(x)\,. \tag{17}$$

Are the solutions of (17) always the same as the solutions of (6)? Of course not. Suppose that $h(s) = 0$, and that s is not a solution of (6). Then $x = s$ is a solution of (17) but not of (6). Therefore the solution set of (6) is not identical with that of (17) so these two equations are not equivalent. But if $x = r$ is a solution of (6), then $x = r$ is also a solution of (17). Therefore the solution set of (6) is contained in the solution set of (17).

Suppose that we divide both sides of equation (6) by $h(x)$ to get

$$\frac{f(x)}{h(x)} = \frac{g(x)}{h(x)}\,. \tag{18}$$

Again the two Equations (6) and (18) need not be equivalent, because in the division some of the roots of (6) may have been canceled by the division in (18).

It is a little difficult to state a rule for multiplication or division of an equation by a function $h(x)$. But a simple warning may be sufficient. Be careful because you may gain or lose some solutions. One should check all of the roots that your work produces, by using them in the original equation to see if they are truly solutions. This will be clarified by

Example 2. Find all of the solutions of

$$\frac{1}{x-3} + \frac{1}{x-11} = 0\,. \tag{19}$$

Solution. Here the fractions are a nuisance. To eliminate the fractions we multiply both sides of (19) by $h(x) = (x-3)(x-11)$, but be careful because $h(x) = 0$ at $x = 3$ and $x = 11$. Multiplication of (19) by $h(x)$ gives

$$\frac{(x-3)(x-11)}{x-3} + \frac{(x-3)(x-11)}{x-11} = 0\,.$$

After cancellation we have

$$(x-11)+(x-3)=0\,,$$
$$2x-14=0\,,$$
$$2x=14\,.$$

If we multiply both sides of this equation by $1/2$ we obtain $x = 7$. Thus the solution of Equation (19) appears to be $x = 7$ but we must be careful. The zeros of $h(x) = (x-3)(x-11)$ are obviously $x = 3$ and $x = 11$, and multiplying by $h(x)$ may have introduced these numbers as solutions. We should check by substituting the suspected roots in the original equation. When we put $x = 3$ in (19) the first term gives $1/0$ and division by zero is undefined (some say it is ∞) so $x = 3$ is NOT a solution of (19). When $x = 11$, the second term in (19) also gives $1/0$, thus $x = 11$ is also NOT a solution. However $x = 7$ gives

$$\frac{1}{7-3}+\frac{1}{7-11}=\frac{1}{4}+\frac{1}{-4}=0\,.$$

Therefore $x = 7$ is the only solution of Equation (19). ►►

In this case multiplication by $h(x)$ did not increase the number of possible solutions.

You may ignore the material between the two horizontal lines.

A student with a vivid imagination might claim that $x = \infty$ (infinity) is also a solution. The argument is as follows. Set $x = M$, a very large number. When this large number is used in equation (19) the left side gives

$$L=\frac{1}{M-3}+\frac{1}{M-11}\,, \tag{20}$$

and if M is very large, then L is very small. Further as M gets larger and larger (approaches infinity) L gets smaller and smaller (approaches zero). Eventually (whatever that means) when $x = \infty$ we have $L = 0$, and hence equation (19) gives $0 + 0 = 0$, a true statement. Therefore $x = \infty$ is a solution of (19).

If this is your argument, you are showing that you have a very creative imagination, and my congratulations to you. What is wrong with the argument? The assumption is that ∞ is a number. If ∞ is a number, it must follow the laws of numbers, and one such law is that the product of two numbers must be uniquely defined. How about the products 2∞ and 3∞? The most reasonable response is they both give infinity. If so, then $2\infty = 3\infty$, and after canceling ∞ (if it is a number) we will find that $2 = 3$. So we cannot accept ∞ as a number, and hence ∞ cannot be a solution of Equation (19). There is a tremendous amount of mathematics that can be done using ∞, but this must wait for a more advanced study. In fact ∞ can be added to the elements of a system in at least two different ways, one way in Projective Geometry, and also in a second (distinctly different) way in the Theory of Complex Variables. But here ∞ is not a number.

Example 3. Find all of the solutions of

$$\frac{x^2}{x^2-4x+3} - \frac{x+2}{x-1} = 0\,. \tag{21}$$

Solution. The first denominator is $(x-1)(x-3)$ and we select this as $h(x)$. When we multiply by $h(x)$ and cancel the common factor $x-1$ in the second term, we obtain

$$x^2 - (x+2)(x-3) = 0\,,$$
$$x^2 - (x^2 - x - 6) = 0\,,$$
$$x + 6 = 0\,.$$

Hence the solution appears to be $x = -6$. The multiplication by $h(x)$ may have introduced new roots, so we substitute $x = 1$ and $x = 3$ in Equation (21) to see if either one is a solution. In both cases the denominator of the first term is zero, so neither $x = 1$ nor $x = 3$ is a solution of (21). If we substitute $x = -6$ in Equation (21) we will find

$$\frac{(-6)^2}{(-6)^2 - 4(-6) + 3} - \frac{-6+2}{-6-1} = 0\,,$$

or

$$\frac{36}{36+24+3} - \frac{-4}{-7} = \frac{36}{63} - \frac{4}{7} = 0\,.$$

But $36/63 = 4/7$, (divide the numerator and the denominator by 9). So this last equation is correct. Thus $x = -6$ is the only solution for Equation (21). ▶▶

Example 4. Find all of the roots of the equation

$$x^2 - 7x + 12 = x^2 - 2x - 3\,. \tag{22}$$

Solution. The simple way is to transpose the right side to the left side, remembering to change signs. This gives

$$x^2 - 7x + 12 - x^2 + 2x + 3 = -5x + 15 = 0\,.$$

Hence $15 = 5x$, or $x = 3$, as the only solution of equation (22). But suppose that the student notices that both sides can be factored so that (22) is equivalent to

$$(x-3)(x-4) = (x-3)(x+1)\,. \tag{23}$$

If we divide both sides of (23) by $h(x) = x - 3$, we obtain

$$x - 4 = x + 1$$

or $0 = 5$. This last (wrong) statement implies that Equation (22) has no solutions, because if there were a solution, then $0 = 5$.

What has happened to the solution $x = 3$? When we divided both sides of (22) by $h(x) = x - 3$, we lost $x = 3$ as a possible solution. In this example the error is easy to spot, but there are many more complicated looking problems, where the error is cleverly disguised and hard to locate. ▶▶

Exercise 2

In Problems 1 through 14 find all solutions of the given equation and check your results by substituting each solution in the original equation. Here, the factoring required is rather simple. We will look at factoring in more detail in Chapter 7 Sections 1 and 2.

1. $x^2 = 3x$.

2. $x^4 = 36x^2$.

3. $10x^2 = 6x^2 + 13x$.

4. $x^2 = 7x - 6$.

5. $2x^3 + 5x^2 + 8x = x^3 + 2x^2 + 6x$.

6. $\dfrac{1}{x-5} + \dfrac{1}{x+7} = 0$.

7. $\dfrac{1}{x-9} + \dfrac{1}{x-5} = 0$.

***8.** $\dfrac{1}{x-a} + \dfrac{1}{x-b} = 0$, where $a \neq b$.

***9.** $\dfrac{1}{x-5} - \dfrac{1}{x+7} = 0$.

10. $\dfrac{2}{x-5} - \dfrac{1}{x+7} = 0$.

11. $\dfrac{x^2}{x^2-4x+3} - \dfrac{x+4}{x-1} = 0$.

12. $\dfrac{x-7}{x-5} - \dfrac{x-5}{x-7} = 0$.

***13.** $\dfrac{x-a}{x-b} - \dfrac{x-b}{x-a} = 0$, where $a \neq b$.

14. $\dfrac{15}{x^2} + \dfrac{8}{x} + 1 = 0$.

3. Linear Equations

An equation that can be put in the form

$$ax + b = 0\,, \qquad a \neq 0\,, \tag{24}$$

is called a *linear equation.* Here a and b are constants and x is the variable. But the definition applies to the form and not the particular letters used. Thus, $cy + d = 0$ is a linear equation in y, and $fs + g = 0$ is a linear equation in s. This same equation is linear in f and it is also linear in g. The equation

$$uv + u - v = 17 \tag{25}$$

is linear in either u or v, but it is not linear in both variables at the same time.

A linear equation is easy to solve. To solve Equation (24) transpose b to the right side and divide by a. The solution is

$$x = \frac{-b}{a}\,, \qquad a \neq 0\,. \tag{26}$$

Example 1. Solve $19x + 41 = 0$.

Solution. We need not memorize Equation (26) because common sense is sufficient. We subtract 41 from both sides of the given equation, and then divide by 19. This will give

$$x = -41/19\,,$$

and since 41 and 19 are both primes this fraction is the solution in lowest terms. ▶▶

Example 2. Solve Equation (25) for u in terms of v.

Solution. The form of the question means that we are to regard u as the variable and all other letters as representing constants. From Equation (25) we have

$$u(v+1) = 17 + v\,,$$

so

$$u = \frac{v+17}{v+1}\,, \qquad v \neq -1\,,$$

and this expresses u in terms of v. Of course we must rule out $v = -1$ because if $v = -1$, then $v + 1 = 0$ and division by 0 is not possible.

If we were asked to solve (25) for v is terms of u we would have

$$v(u-1) = 17 - u\,,$$

so

$$v = \frac{-u+17}{u-1}\,, \quad u \neq 1\,. \qquad ▶▶$$

Example 3. Is the equation

$$abcd + bcde + cdea + deab + eabc = 13 \tag{27}$$

linear in b? If so solve the equation for b.

Solution. The early letters in the alphabet usually denote constants and the later letters such as x, y, and z denote variables. But this is a custom and not a strict law. So a, b, c, d, and e, may be regarded as variables if the problem under consideration demands it. Now b occurs to the first power or doesn't appear in each term of (27) so the equation is linear in b. In (27) we transpose the terms that do not contain b to the right side, factor b from the remaining terms. Then we have

$$b(acd + cde + dea + eac) = 13 - cdea\,,$$

and hence

$$b = \frac{13 - cdea}{acd + cde + dea + eac}\,.$$

whenever the denominator is not zero. ▶▶

Exercise 3

In Problems 1 through 15 solve the given equation for x.

1. $7x + 9 = 5x + 3$.

2. $19x + 41 = 13x + 31$.

3. $\frac{1}{2}x + \frac{1}{3} = \frac{1}{4}$.

4. $\frac{1}{5}x - \frac{1}{4} = \frac{1}{3}$.

5. $5(x-1)^2 = 5x^2 - 14x - 17$.

6. $(2x+3)^2 = (2x-5)^2$.

7. $(3x+5)^2 = (3x-11)^2$.

8. $5x^2+17x+42 = 3x(x-5)+2x(x-3)$.

9. $\dfrac{5x}{x-8} + \dfrac{2x}{x+2} = 7$.

10. $\dfrac{5x}{x-8} - \dfrac{2x}{x+2} = 3$.

11. $\dfrac{7x}{x-1} + \dfrac{5x}{x-4} = 12$.

12. $\dfrac{7x}{x-1} - \dfrac{5x}{x-4} = 2$.

13. $\dfrac{2x-1}{x-2} + \dfrac{3x-2}{x-1} = 5$.

14. $\dfrac{2x-3}{x-2} + \dfrac{3x-2}{x-3} = 5$.

15. $\dfrac{2x-2}{x-4} + \dfrac{3x-4}{x-2} = 5$.

16. Solve $(2x-a)/(x-b) + (3x-b)/(x-a) = 5$, for x in terms of a and b. Observe that this problem contains Problems 13, 14, and 15 when we select a and b properly. Thus by solving Problem 16 we solve those three problems simultaneously together with infinitely many other problems as well. What values of a and b will give Problems 13, 14, and 15?

In Problems 17 through 25 solve the given equation for y.

17. $xy + yz + zx = 2$.

***18.** $\dfrac{xy + yz + zx}{x + 2y + 3z} = 2$.

19. $\dfrac{x}{y} + \dfrac{z}{x} = xz$.

20. $x^2 + (y-p)^2 = (y+p)^2$. This is the equation of a parabola when it is in a standard position and the distance from the focus to the vertex is $2p$.

21. $\dfrac{x^2}{a^2} + \dfrac{y^2}{b^2} = 1$. Two solutions. This is the equation of an ellipse in standard position.

22. $(y-a)^2 + (y-b)^2 + (y-c)^2 = 3y^2$.

23. $(y-ax)^2 + (y-bx)^2 + (y-cx)^2 = 3y^2$.

24. $(x-a)^2 + (y-b)^2 = r^2$. Two solutions. This is the equation of a circle with center at (a,b) and radius r.

25. $(x-a)^2 + (y-b)^2 + (z-c)^2 = r^2$. This is the equation of a sphere with center at (a,b,c) and radius r.

4. Applications of Linear Equations

The recent meeting of an American spacecraft and a Russian spacecraft while in orbit and traveling at a high velocity, was a real triumph of engineering and technology. But it was an even *greater triumph* for mathematics which was necessary for the computation of the exact meeting place and the exact time. One does not begin the study of applied mathematics with such horribly difficult problems. You must start with extremely simple problems and as your knowledge of mathematics increases you can attack more difficult and more interesting problems. In this section we look at the simplest types of applied problem. These problems may seem childish and uninteresting to you, but remember that by tackling and solving very elementary problems you are building the strength to go further in mathematics and attack more difficult and more important problems. Further, working the simple problems can be regarded as fun.

Applied problems are often described by words, and it is the duty of the mathematician to translate the words into equations and use mathematics to arrive at the answer to the question that is stated (or implied) in the word description of the problem. Many years ago such problems were called *story problems.* Later, they became *word problems.* Today, the process of going from words to equations is called *modeling.* The idea is that the equations do not truly describe the situation, because some unimportant items have been ignored. As an example, consider the statement "Alvin runs at the rate of 20 feet per second (20 ft/sec)". Now the statement cannot be exactly true because Alvin's speed is usually slower at the start, possibly much higher than 20 ft/sec in the middle of a race, and again slower at the end due to fatigue. When we ignore such variations in the data while translating the words into equations we are said to be *making a model.* When we solve the equations, we are really solving the model, and it is possible that the solution of the model, is quite different from the solution supplied by nature in the real world.

Many problems are so difficult that no reasonable model is available at this time. Perhaps the outstanding example is the problem "How do we achieve peace in the world in our time?" In 1938 the English prime minister, Neville Chamberlain, announced that he had solved this problem, but the events that followed (World War II) showed that he was ridiculous and completely wrong.

For those word problems that are amenable to a mathematical treatment, the following outline of steps may be helpful.

(A) Search for those items in the problem that can be measured.

(B) Introduce letters (variables) as needed to represent the measures of those items. It is a good idea to write the precise meaning of each symbol. This will help you to concentrate your full attention on what you are trying to do.

(C) Read the problem carefully for information that gives an equation (or equations) relating those variables. Sometimes a drawing will be helpful.

(D) Be careful to observe that some of the information given in the problem is completely useless.

(E) If there are several equations and several variables, use some of the equations (properly) to transform one of the equations into an equation that involves only one variable.

(F) Solve the equation in one variable (if possible).

(G) Use the result from (F) to find the values of the other variables.
(H) Check that your solution satisfies the conditions stated in the problem.

There are many nice problems that lead to linear equations in steps (E) and (F). In this chapter we present a reasonable selection of such problems.

Example 1. A Transvalley airplane leaves Wildroot airport at exactly 12:07 P.M. flying due north. Two hours later a Quick Serve plane departs from the same airport flying in the same direction. The second plane overtakes the first one at 7:07 P.M. just as dinner is being served on both planes. If the Quick Serve plane flies 60 miles/hour faster than the Transvalley plane, what is the speed of each plane? How far are they from Wildroot airport when they meet?

Solution. We follow the outline suggested.

(A) Of course the names of the airlines and the name of the airport give useless information and should be ignored.
(B) Introduce letters. Select the letters you prefer. I like these.
Let x = rate of travel (speed in miles/hour) of the Transvalley plane.
Let y = rate of travel (speed in miles/hour) of the Quick Serve plane.
Let d_1 = distance traveled by the Transvalley plane.
Let d_2 = distance traveled by the Quick Serve plane.
(C) Of course when they meet we have $d_1 = d_2$ but let us reserve that information until later. Also the problem states that

$$y = x + 60\,. \tag{28}$$

We assume that the speed of each plane is constant throughout the flight. Whenever the speed is constant we have the fundamental relation

$$\boxed{\text{distance traveled} = (\text{rate of travel})(\text{time of travel})}$$

or

$$\boxed{d = rt\,.} \tag{29}$$

(In fact, Equation (29) is the definition of rate, since $r = d/t$.) For the Transvalley plane, Equation (29) gives $d_1 = xt_1$ and for the Quick Serve plane $d_2 = yt_2$. But reading the problem carefully we see that when the planes meet $d_1 = d_2$ while $t_1 = 7$ hours and $t_2 = 5$ hours. Therefore $d_1 = 7x = d_2 = 5y$ or $7x = 5y$.
(E) The equation $7x = 5y$ in two variables does not give us enough information to solve the problem. But we can use Equation (28) which gives $y = x + 60$, to eliminate the variable y in $7x = 5y$. Replacing y in this equation by $x + 60$ gives

$$7x = 5(x + 60)\,,$$

a linear equation, that is easy to solve.
(F) Thus, $7x - 5x = 5(60) = 300$. Hence $2x = 300$ and thus $x = 150$ miles/hour.

(G) Once we have $x = 150$, it is easy to find the other variables. From (28), we see that $y = x + 60 = 150 + 60 = 210$ miles/hour. The distance traveled when they meet is

$$7(150) = 5(210) = 1{,}050 \text{ miles}.$$ ►►

Example 2. An industrial chemist has 96 gallons of 70% sulfuric acid and wishes to add enough water so that the resultant mixture will be 40% sulfuric acid. How many gallons of water should be added?

Solution. Let x be the number of gallons of water to be added. The amount of pure sulfuric acid is the same both before and after the addition of the water. By the meaning of the word "percent" we have the amount of pure sulfuric acid *BEFORE* adding water is $0.70(96)$. The amount of pure sulfuric acid *AFTER* adding water is $0.40(96 + x)$.
Since the amount of pure sulfuric acid is the same during the process, this gives the equation

$$0.70(96) = 0.40(96 + x)\,. \tag{30}$$

In problems of this type it often helps to multiply both sides of an equation by a suitable number to eliminate fractions. For percent type problems the multiplier 100 usually works. In this case the factor 10 will be enough to eliminate fractions. When we multiply both sides of Equation (30) by 10 we obtain

$$7(96) = 4(96) + 4x$$

or

$$4x = 7(96) - 4(96) = (7 - 4)(96) = 3(96) = 288\,.$$

Therefore $x = 288/4 = 72$. The chemist should add 72 gallons of water. ►►

Example 3. One pump working steadily will fill a reservoir in 30 days. A second pump will do the same job in 20 days. If both pumps are used together how long will it take to fill the reservoir?

Solution. This problem seems to be very difficult because we have no information about the capacity of the reservior in cubic yards, or in gallons. We do not know what units of measure to use and so it seems as though we have hit a barrier. There is a simple way around this difficulty. We let the reservoir itself be the unit of volume. If one pump fills one reservoir in 30 days, then the rate of pumping is $r_1 = 1/30$ res/day (reservoirs per day). For the second pump $r_2 = 1/20$ res/day. Thus, when both pumps are working together, the combined rate (which we denote by R) is given by

$$R = \frac{1}{30} + \frac{1}{20} \quad \text{reservoirs/day}\,. \tag{31}$$

How do we know that we can add the rates r_1 and r_2 as we did in Equation (31)? Here the student might review the earlier material on modeling. One must look more deeply into the real nature of the quantities involved. Suppose that r_1 and r_2 refer to the speed of two runners. Whether they are racing each other, or are partners in a relay race, the equation

$R = r_1 + r_2$ makes no sense and cannot be used for their joint rate. On the other hand, if the two people or two machines are working together on a job, then one can present a good argument for adding their rates as we did in Equation (31). The model is not perfect. For example if two painters are working together to paint a house, they might interfere with each other and the joint rate might be less than the sum. Or by talking to each other or racing, the joint rate could be greater than the sum. From here on, such minor factors will be ignored, and wherever it seems reasonable we will use the equation $R = r_1 + r_2$ or its extension to more than two rates.

To return to Example 3, in place of the usual $tr = d$ for a moving object we now have

$$\boxed{\text{(time)(rate of accomplishment)} = \text{(fraction of the job completed)}.}$$

In this case the time t to fill the reservoir is determined by

$$t\left(\frac{1}{30} + \frac{1}{20}\right) = 1\,, \quad \text{(1, because we wish to fill the reservior)}\,.$$

Now, $1/30 + 1/20 = 50/600$ or $5/60$. Therefore $t(50/600) = 1$ or

$$t = 600/50 = 12 \quad \text{days}\,.$$

This makes sense because working together the time required by the pumps should be less than the 30 or 20 days required by the individual pumps working alone. ▶▶

Exercise 4

1. Suppose that the two planes in Example 1 fly in opposite directions and that at 8:07 P.M. they are 2,550 miles apart. Find the speed of each plane if the Quick Serve plane flies 40 miles/hour faster than the Transvalley plane.

2. A Transvalley plane leaves Sirloin City for Steakville 1,000 miles away at 7:43 A.M. at 8:43 A.M. a Quick Serve plane leaves Steakville for Sirloin City. If the Quick Serve plane travels 70 miles/hour faster than the Transvalley plane and passes over that plane at 10:43 A.M., find the speed of each plane.

3. A chemist has 26 gallons of 60% nitric acid. How much water should he add to obtain (a) a 40% nitric acid mixture, (b) a 20% nitric acid mixture? Is the answer to part (b) twice the answer to part (a)?

4. A tank car is partially full of 55% hydrochloric acid. After 9,000 gallons of water were added a test showed that the mixture was 46% hydrochloric acid. How many gallons of 55% acid did the tank car contain initially?

5. Arvin can mow the lawn in 6 hours with an old fashioned push type mower, and Bozo can do the same lawn in 2 hours with a power mower. How long will it take them to mow the lawn if they work together using both mowers?

6. Alice can paint her room is 5 hours 30 minutes. Janet can paint the same room in 4 hours 30 minutes. If they work together how long will it take them to paint the room?

7. The sum of three consecutive integers is 249. Find the first one of these three integers.

8. The sum of four consecutive integers is 274. Find the first one of these four integers.

9. The sum of four consecutive odd integers is 1,096. Find the first one of these four integers.

10. The sum of six consecutive odd integers is 492. Find the first one.

11. Find two numbers which differ by 4 and such that their squares differ by 200.

12. Find two numbers which differ by 6 and such that their squares differ by 267.

13. Two numbers have the sum 513, and one of the numbers is 6 times the other number. Find the numbers.

14. Three numbers have the sum 765. The second number is twice the first and the third number is three times the second. Find the three numbers.

15. Zeno has found that if he increases his average speed from 50 miles/hour to 60 miles/hour he can save 15 minutes in driving from Wildwood to Lakeview. Find the distance from Wildwood to Lakeview.

16. One hour and 15 minutes after Alice left home on her bicycle Betty went after her in a car which averages 25 miles/hour faster than the bicycle. After driving 45 minutes Betty caught up with Alice. Find the speed of the car.

17. An airplane with a ground speed of 355 miles/hour finds that against the wind a certain trip requires 6 hours, but the return trip with the wind takes only 4 hours. Assuming that the wind is steady, find the speed of the wind.

18. Going home from work Zeno speeds 15 minutes going steadily through town traffic, and when he hits route I-75 he increases his average speed by 30 miles/hour. If he spends 10 minutes on I-75 and the total distance from work to home (just at the I-75 exit) is 15 miles, find his speed on I-75.

19. Zunny has $10,000 invested, some at 6% and some at 4%. The total income gives her an average yield of 5.4%. How much was invested at 6%?

20. Abby has invested a total of $8,000, some at 4% and the rest at 7%. If the 4% investment yields $155 more each year than the 7% investment, how much was invested at 4%?

21. A rectangular lot is twice as long as it is wide. A road 20 feet wide through the lot parallel to the long side will use 700 square feet more of the lot than a road 30 feet wide parallel to the short side. Find the dimensions of the lot.

22. A retired professor sold her mansion. With the proceeds she bought a smaller house (for herself) at one-half the price of the mansion and a second house for $10,000 more than one-third the price of the mansion. She had $24,000 left which she spent on a fancy car. How much did she pay for the second house?

23. Arvin, Bozo, and Corky together earn $630 a week. Arvin's weekly wage is $120 less than twice Bozo's, while Corky earns $90 per week more than two-thirds Bozo's salary. Who has the biggest income and how much is it?

24. Arvin, Bozo, and Corky together bought a used car. Arvin put in $300 more than one-fifth of the cost of the car. Bozo contributed $90 more than Arvin. Corky's share was $200 less than one-fourth the cost of the car. What did the car cost?

25. Alice can paint a house in 36 hours. Cora can paint the same house in 24 hours. They started working together but after 12 hours (spread over two days) Alice became discouraged and quit. How much time was necessary for Cora to finish the job?

26. Barney and Corrigan together can build a brick wall in 12 hours. If Corrigan works alone he can do the job in 21 hours. How long would it take for Barney to build the wall by himself?

27. My daughter is eleven years older than my son. In 3 years she will be twice as old as my son. How old are they now?

28. Arlene is five times as old as her daughter. In 18 years Arlene will only be twice as old as her daughter. How old is Arlene now?

29. Arvin, Bozo, and Corky working separately can pick all the oranges in a small grove in 36, 27, and 18 days respectively. If Arvin and Bozo start together and Corky joins them two days later, how many days will it take to clear the grove of oranges?

30. When the boys from Problem 29 moved to a second grove (same size) Arvin started working immediately, Bozo joined him 3 days later and after another 3 days Corky started picking oranges too. How many days did Arvin work before the second grove was harvested?

31. Two cars racing on an oval track 4 miles long average 120 miles/hour and 105 miles/hour respectively. If they start together, how long will it take for the first car to be one lap ahead of the second car?

32. At 4:00 P.M. two criminals in a red Cadillac enter I-75 and travel north at 65 miles/hour. Twenty minutes later a policeman enters I-75 at the same place and chases them at 75 miles/hour. At what time does the policeman overtake the criminals?

33. Do Problem 32 if there are three criminals in a light blue Chevrolet that travels 60 miles/hour, enters I-75 at 1:00 P.M. and the policeman enters I-75 at the same place a half hour later and drives 80 miles/hour.

---

5. Odd and Even Functions

Of course we need

Definition 2. A function $f(x)$ is called an *even function* if

$$f(x) = f(-x) \tag{32}$$

whenever x and $-x$ are both in the domain of $f(x)$. The function is called an *odd function* if

$$f(x) = -f(-x) \tag{33}$$

whenever x and $-x$ are both in the domain of $f(x)$.

Example 1. Is $f(x) = x^m$ an even function or an odd function when m is an integer?

Solution. If m is an even integer, then x^m is an even function. If m is an odd integer, then x^m is an odd function. ▶▶

Indeed this is the reason the functions are given the titles "even" and "odd"

Example 2. Is the function $g(x) = \sqrt{x-7}$ an even function or an odd function?

Solution. Since the domain of $g(x)$ is the ray $[7, \infty)$ there is no x such that both x and $-x$ are in the domain of $g(x)$. Thus this function satisfies the definition vacuously (there is nothing to test). Thus if we follow the definition strictly, then $g(x)$ is both an even function and an odd function.

This ridiculous situation is introduced to show the difficulty of writing a definition accurately. The definition should be worded to restrict the test to those functions for which the domain contains some points and contains $-x$ whenever the domain contains x. Thus an accurate definition may become so wordy and complex that the student has trouble understanding the essential concept. For this reason many authors will present a simplified (and incorrect) statement (as we have done) to make it easy for the student to grasp the idea.

Example 3. Is $H(x) = 7x^4 - 5x^2 - \pi^3$ an odd function, an even function, or neither?

Solution. Since $H(-x) = 7(-x)^4 - 5(-x)^2 - \pi^3 = 7x^4 - 5x^2 - \pi^3 = H(x)$, the function is an even function. ▶▶

Exercise 5

(A)

In Problems 1 through 6 determine whether the given function is an odd function, an even function, or neither.

1. $f(x) = x^{10} - 5x^6 - 7x^4 - 11x^2$.

2. $g(x) = x^7 + 6x^5 + 12x^2$.

3. $F(y) = y^3 + 1/y^5$.

4. $G(x) = \dfrac{x^2}{5x^4 - 43}$.

5. $H(z) = \dfrac{z^3}{3z^9 + 11z^5 - 77z}$.

6. $J(x) = \dfrac{x^2 + 7x^4 - 21x^6}{4x + 102x^3}$.

In Problems 7, 8, and 9 form the new function

$$g(x) = f(x) + f(-x)$$

using the $f(x)$ given. Is $g(x)$ an even function, an odd function, or neither?

7. $f(x) = 5x^2 - 2x^3$.

8. $f(x) = 3x^5 - 4x^4 + 5x^3 + 6x^2$.

9. $f(x) = \sqrt{13}x^{10} - \sqrt{6}x^7$.

10. Form the new function $h(x) = f(x) - f(-x)$ using the function $f(x)$ from (A) Problem 7, (B) Problem 8, and (C) Problem 9. Is $h(x)$ an even function, an odd function, or neither?

***11.** Prove that if $f(x)$ is any function defined on a suitable domain (see the discussion after Definition 2) and we set

$$g(x) = f(x) + f(-x),$$

then $g(x)$ is an even function.

***12.** Prove that if $f(x)$ is any function defined on a suitable domain and we set

$$h(x) = f(x) - f(-x),$$

then $h(x)$ is an odd function.

****13.** Prove that if $f(x)$ is any function defined on a suitable domain, then $f(x)$ can be written as the sum of an odd function and an even function. Hint: Use the results from Problems 11 and 12. You will need to divide by 2 at some time.

Chapter 7

Quadratic Equations

1. Factoring Quadratic Expressions

If an equation can be put in the form

$$ax^2 + bx + c = 0\,, \tag{1}$$

the left side is called a *quadratic expression*, and Equation (1) is called a *quadratic equation*.

In this section we will learn to solve such equations by factoring. We have already seen that if the left side of Equation (1) can be factored, then the equation is easy to solve. See Chapter 5, Section 1. Because this material is so important, we review it again here (with a slight change in some of the symbols).

We suppose first that $a = 1$ in Equation (1), so that our quadratic equation has the form

$$x^2 + bx + c = 0\,. \tag{2}$$

If the left side can be factored so that

$$x^2 + bx + c = (x + r)(x + s) = 0\,, \tag{3}$$

then:

$$\text{Either} \quad x + r = 0 \qquad\qquad \text{or} \quad x + s = 0 \tag{4}$$
$$x = -r\,, \qquad\qquad\qquad x = -s\,.$$

So $r_1 = -r$ and $r_2 = -s$ are the roots (solutions) of Equation (2), and they are the only solutions. (Watch out for the negative signs.)

If we want to factor the quadratic polynomial $x^2 + bx + c$, we must find numbers r and s such that $r + s = b$ and $rs = c$. Thus the roots satisfy the equations

$$r_1 + r_2 = -(r + s) = -b\,, \quad \text{and} \quad (r_1)(r_2) = (-r)(-s) = c\,. \tag{5}$$

We summarize these results in

Theorem 1. *If $Q(x) = x^2 + bx + c = (x + r)(x + s)$, then the equation $Q(x) = 0$ has the roots (solutions) $r_1 = -r$, and $r_2 = -s$. Conversely, if r_1 and r_2 are the solutions of the quadratic equation $Q(x) = 0$, then $r_1 + r_2 = -b$, and $r_1 r_2 = c$ in Equation (3).*

Example 1. Solve each of the following equations

$$Q_1(x) = x^2 + 8x + 15 = 0\,,$$
$$Q_2(x) = x^2 + 2x - 15 = 0\,,$$
$$Q_3(x) = x^2 - 2x - 15 = 0\,,$$

and

$$Q_4(x) = x^2 - 8x + 15 = 0\,.$$

Solution. In the first quadratic $Q_1(x)$ we are looking for r and s such that $rs = 15$ and $r + s = 8$, (see Equation (5)). Obviously, $r = 5$ and $s = 3$ will do very nicely, since the sum is 8 and the product is 15. Therefore for $Q_1(x)$

$$Q_1(x) = x^2 + 8x + 15 = (x+3)(x+5) = 0\,,$$

and hence the solutions are $x = -3$, and $x = -5$, (see Equation (4)).

For $Q_2(x)$, keep in mind that the numbers r and s in (5) may be positive or negative. Since $5 - 3 = 2$ and $5(-3) = -15$, it follows that $r = 5$ and $s = -3$, and hence

$$Q_2(x) = x^2 + 2x - 15 = (x+5)(x-3)\,.$$

Therefore, the solutions are $x = -5$ and $x = 3$, (the negative of 5 and the negative of -3).

For $Q_3(x)$, we have $-5 + 3 = -2$ and $(-5)3 = -15$. Hence $r = -5$ and $s = 3$. Therefore,

$$Q_3(x) = x^2 - 2x - 15 = (x-5)(x+3)\,,$$

and hence the solutions are $x = -r = -(-5) = 5$, and $x = -s = -3$.

Finally, for $Q_4(x)$ we have $-5 - 3 = -8$ and $(-5)(-3) = 15$. Thus

$$Q_4(x) = x^2 - 8x + 15 = (x-5)(x-3)\,,$$

and hence the solutions are $x = 5$, and $x = 3$. ▶▶

With lots of practice, the student should be able to do this type of factoring "in his head" writing the product without any intermediate hand work on paper. Then, to find the roots of the equation one must alter the sign of r and s.

Suppose that the leading coefficient in our quadratic is not 1, so that the quadratic that we wish to factor is $ax^2 + bx + c$. If we expand the middle term in

$$ax^2 + bx + c = (px + r)(qx + s) = pqx^2 + (ps + rq)x + rs\,, \tag{6}$$

we will obtain the expression on the extreme right. Thus Equation (6) will become an identity if the corresponding coefficients are equal. By this we mean that

$$\begin{cases} a = pq\,, & \text{the coefficients of } x^2 \text{ are the same}\,, \\ b = ps + rq\,, & \text{the coefficients of } x \text{ are the same}\,, \\ c = rs\,. & \text{the constant term is the same}\,, \end{cases} \tag{7}$$

for the two polynomials at the two ends of Equation (6).

Thus to factor $ax^2 + bx + c$ we must solve the system of equations (7) for p and q. This looks much harder than it is. Using only integers for p and q, there are only a finite number of choices, and it is usually easy to select the correct p and q.

Example 2. Solve the equation

$$6x^2 + 11x - 7 = 0 \tag{8}$$

and check that your solutions really give 0 on the left side of (8).

Solution. Although it seems as though we should use the equation set (7) to solve this problem, we are actually much better off if we ignore those miserable looking equations and return to the source. We want an identity of the form

$$6x^2 + 11x - 7 = (px + r)(qx + s)\,, \tag{9}$$

where p, q, r, and s are 4 integers. Clearly we must have $pq = 6$. Now there are only a finite number of pairs of integers that work. These are $(1, 6)$, $(-1, -6)$, $(2, 3)$, and $(-2, -3)$. You can ignore the pair $(6, 1)$ because one can always change the order of multiplication on the right side of (9). So essentially there are at most 4 choices for (p, q). For r and s the product must be -7 so we have only to consider the pairs $(7, -1)$, $(-1, 7)$, $(-7, 1)$, and $(1, -7)$. Having listed these, it is a simple matter to look at the middle term of Equation (7)

$$11 = ps + rq$$

and select $p = 2$, $q = 3$, $r = -1$, and $s = 7$ to give $ps + rq = 11$. Hence

$$6x^2 + 11x - 7 = (2x - 1)(3x + 7)\,.$$

This product is zero if and only if

Either $2x - 1 = 0\,,$	or $3x + 7 = 0\,,$
$2x = 1\,,$	$3x = -7$
$x = \dfrac{1}{2}\,,$	$x = \dfrac{-7}{3}\,.$

To check the solution $x = -7/3$, we substitute this number in the right side of (8) and obtain

$$\begin{aligned} 6x^2 + 11x - 7 &= 6\left(\frac{49}{9}\right) + 11\left(\frac{-7}{3}\right) - 7 \\ &= \frac{2(49) - 77 - 21}{3} = \frac{98 - 98}{3} = 0\,. \end{aligned}$$

We leave it for the reader to check the solution $x = 1/2$. ►►

Example 3. Find all the real roots of $G(z) = 4z^4 + 59z^2 - 15 = 0$.

Solution. Wait a minute, you scream. $G(z)$ is not a quadratic and this chapter is on quadratic equations. But it really is a quadratic in disguise. Because if we put $z^2 = x$ so

$z^4 = x^2$ then $G = 0$ if and only if $4x^2 + 59x - 15 = 0$, and this latter is a quadratic equation. (We can say that $G(z)$ is a quadratic expression in z^2.)

Returning to our original $G(z)$ and observing that the terms z^3 and z are missing in $G(z)$ we look for a product of the form

$$4z^4 + 59z^2 - 15 = (pz^2 + r)(qz^2 + s)\,.$$

Without bothering to examine the equation set (7) it is clear that we must have $pq = 4$ and $rs = -15$. If we are restricted to integers, the only pairs possible for (r, s) are $(15, -1)$, $(-15, 1)$, $(1, -15)$, $(-1, 15)$, $(5, -3)$, $(-5, 3)$, $(3, -5)$, and $(-3, 5)$. A similar analysis shows that there are only six possible choices for the pair (p, q). We leave it for the reader to list these 6 pairs. If we try all $6(8) = 48$ possible pairs we will soon find that

$$G(z) = (4z^2 - 1)(z^2 + 15) = (2z - 1)(2z + 1)(z^2 + 15)\,. \tag{10}$$

But we do not need to try all 48 combinations of p, q, r, and s. The fact that the coefficient of z^2 is 59 tells us loudly that we will need the product of 4 and 15, and this greatly reduces the number of choices we can make.

The first two factors on the right side of (10) give the solutions $z = +1/2$ and $z = -1/2$. If we are looking only for real roots, the last factor $z^2 + 15$ in Equation (10) is always positive for real z and hence does not yield any real roots. Roots that are complex numbers will be considered in the next section. ►►

Example 4. Solve the equation

$$x^2 + x - 3 = 0\,. \tag{11}$$

Solution. No matter how hard we try, we cannot find *integers* r and s such that $x^2+x-3 = (x + r)(x + s)$. Thus, with our present methods we cannot solve this problem. ►►

In fact if we select at random a quadratic equation with integer coefficients, probably we will not be able to factor it using only integer coefficients in the factors. The curious reader might try this experiment using the roll of dice, or drawing cards to determine the coefficients in the quadratic equation.

What shall we do? This will be covered in the next section.

When a polynomial *cannot* be factored with coefficients in a certain set, the professional mathematicians say that the polynomial is *irreducible over that set* (not factorable over the set). This is the case in Example 4 where the set is the set of integers.

Exercise 1

In Problems 1 through 24 solve the given equation.

1. $x^2 + 2x - 3 = 0.$

2. $x^2 - 7x + 6 = 0.$

3. $x^2 - 2x - 15 = 0.$

4. $y^2 + 5y - 24 = 0.$

5. $y^2 + 2y - 24 = 0.$

6. $x^2 - 16x + 28 = 0.$

7. $z^2 - 5z - 36 = 0.$

8. $z^2 - 6z - 40 = 0.$

9. $a^2 - 13a - 48 = 0.$

10. $b^2 + 32b + 60 = 0.$

11. $2x^2 - 7x + 6 = 0.$

12. $2x^2 - 5x - 18 = 0.$

13. $3x^2 + 11x - 4 = 0.$

14. $6t^2 - 7t + 2 = 0.$

15. $3t^2 - 7t - 6 = 0.$

16. $5x + 4 = 1/x.$

17. $6x - 1 = 12/x.$

18. $6y + 7 = 3/y.$

19. $6x^2 = 12 + x.$

20. $2x = 1 - 3x^2.$

21. $x^4 - 13x^2 + 36 = 0.$

22. $x^4 - 21x^2 + 80 = 0.$

23. $x^4 - 3x^2 + 2 = 0.$

24. $x^4 + 18 = 9x^2.$

In Problems 25 through 28, find all the *real* roots of the given equation.

25. $x^4 - 1 = 0.$

26. $x^4 + x^2 = 20.$

27. $x^4 - x^2 = 20.$

28. $x^4 + 9x^2 + 20 = 0.$

In Problems 29 through 38 find an equation whose roots r_1 and r_2 have the given property. Hint: Recall from Theorem 1 and Equation (5) that $r_1 = -r$ and $r_2 = -s$. Thus in x^2+bx+c we have $b = r + s = -(r_1 + r_2)$, and $c = rs = (-r_1)(-r_2) = r_1r_2$.

29. $r_1 + r_2 = 19, \quad r_1r_2 = 60.$

30. $r_1 + r_2 = -12\pi, \quad r_1r_2 = 11\pi^2.$

31. $r_1 + r_2 = 2, \quad r_1r_2 = -48.$

32. $r_1 + r_2 = -3, \quad r_1r_2 = -180.$

33. $r_1 = 7, \quad r_2 = -11.$

34. $r_1 = 3, \quad r_2 = -4.$

35. $r_1 = 1/2, \quad r_2 = 1/3.$

36. $r_1 = 1/3, \quad r_2 = -1/4.$

37. $r_1 = 2/5, \quad r_2 = -3/5.$

38. $r_1 = -7/2, \quad r_2 = 8/3.$

39. If $f(x) = x^2 - 5x + 17$, find all x such that $f(x) = f(-2x)$.

40. If $f(x) = x^2 + 8x + 13$, find all x such that $f(3x) = f(-x)$.

41. Solve $6 = 7y^2 - y^4$. Hint: First let $y^2 = x$ and solve for x. Then find y from $y^2 = x$.

42. Solve $5 = 6u^2 - u^4$.

43. Solve $6 = \dfrac{5}{y-1} - \dfrac{1}{(y-1)^2}$. Hint: let $x = \dfrac{1}{y-1}$.

44. Solve $4 = \dfrac{5}{q-2} - \dfrac{1}{(q-2)^2}$.

2. The Quadratic Formula

We return for the moment to Example 4 of Section 1, where we were asked to solve the equation

$$x^2 + x - 3 = 0\,. \tag{11}$$

We recall that we were not able to factor the left side of (11) using only integer coefficients. Does that equation have solutions? The answer is yes, as we will see in a moment. But we will do much much better than merely solve Equation (11). We will derive (prove) a remarkable formula that *solves all quadratic equations at the same time.* Think about it for a moment. In place of trying to factor the quadratic expression in Equation (11), we can use this formula to solve *any quadratic equation* no matter how complicated the coefficients might be. Here is the derivation of this remarkable formula.

Every quadratic equation can be put in the form

$$ax^2 + bx + c = 0\,. \tag{12}$$

Now $a \neq 0$, for if $a = 0$, then Equation (12) $ax^2 + bx + c = 0$, becomes $bx + c = 0$ and the equation is not a quadratic. We can divide Equation (12) by a, the leading coefficient, and when we do this we obtain

$$x^2 + \frac{b}{a}x + \frac{c}{a} = 0\,.$$

Subtract c/a from both sides.

$$x^2 + \frac{b}{a}x = -\frac{c}{a}\,. \tag{13}$$

We now add to both sides a magic number that makes the left side a perfect square. This magic number is $b^2/4a^2$.

$$x^2 + \frac{b}{a}x + \frac{b^2}{4a^2} = \frac{-c}{a} + \frac{b^2}{4a^2}\,. \tag{14}$$

The left side is a perfect square. It is the square of $x + \frac{b}{2a}$. Square this, and see for yourself.

$$\left(x + \frac{b}{2a}\right)^2 = \frac{-4ac}{4a^2} + \frac{b^2}{4a^2}\,. \tag{15}$$

On the right side we made $4a^2$ a common denominator.

Next, we rearrange the terms on the right side and combine the terms with the same denominator.

$$\left(x + \frac{b}{2a}\right)^2 = \frac{b^2 - 4ac}{4a^2}\,. \tag{16}$$

Next, we take the square root of both sides of (16).

We must insert the selection $\pm$ sign because (for example) both 5 and -5 give 25 when we square them.

$$x + \frac{b}{2a} = \pm\sqrt{\frac{b^2 - 4ac}{4a^2}}.$$

The square root of the denominator on the right side is $2a$.

$$x + \frac{b}{2a} = \frac{\sqrt{b^2 - 4ac}}{2a}. \tag{17}$$

Finally, subtract $b/2a$ from both sides of (17) and observe the common denominator $2a$ on the right side.

$$\boxed{x = \frac{-b \pm \sqrt{b^2 - 4ac}}{2a}.} \tag{18}$$

This is it! Equation (18) is the magic formula that solves all quadratic equations at the same time.

We have proved that if the quadratic equation $ax^2 + bx + c = 0$ has a solution, then it is given by (18). But *all the steps can be reversed.* Therefore, if x is given by (18), it *is a solution of (12)*. We have proved

Theorem 2. *All solutions of the quadratic equation*

$$Q(x) = ax^2 + bx + c = 0\,, \quad a \neq 0\,, \tag{12}$$

are given by the formula

$$\boxed{x = \frac{-b \pm \sqrt{b^2 - 4ac}}{2a}.} \tag{18}$$

Conversely, if x is given by Equation (18) *then x is a solution of the quadratic Equation* (12).

The right side of (18) is called the *quadratic formula.*

You should memorize Equation (18), the quadratic formula.

This very important formula deserves an exhaustive discussion. But before giving any of the history and philosophy, let us see how it is used.

Example 1. Use the quadratic formula to solve $x^2 + x - 3 = 0$.

Solution. Matching this equation with Equation (12) we see that for this equation: $a = 1$, $b = 1$, and $c = -3$. From (18) we have

$$x = \frac{-b \pm \sqrt{b^2 - 4ac}}{2a} = \frac{-1 \pm \sqrt{(1)^2 - 4(1)(-3)}}{2(1)} = \frac{-1 \pm \sqrt{13}}{2}.$$

Hence there are exactly two different solutions of $x^2 + x - 3 = 0$. These are

$$r_1 = \frac{-1 - \sqrt{13}}{2}\,, \quad \text{and} \quad r_2 = \frac{-1 + \sqrt{13}}{2}. \tag{19}$$

We can check our work, and at the same time check our theory, if we recall Theorem 1. This states that $r_1 + r_2 = -b$, and $r_1 r_2 = c$. In this case the roots given in (19) must have the sum $-b = -1$ and the product $c = -3$. We leave it for the reader to show that this is really the case for the two roots given by Equation (19). ►►

Look carefully at the proof of Theorem 2. Probably this is the first bit of *real algebra* that you have seen. Up to now you have been concerned with the basic relations and drill work in using them (adding fractions, factoring, cancellation). Now you have proved a theorem that solves infinitely many problems all at the same time. For the first time you have seen a demonstration of the powerful thought processes of imaginative mathematicians. You should memorize the proof because this will help you to understand this great achievement. Of course mathematics contains many more theorems, more complicated, more powerful, and more beautiful, but Theorem 2 is a good place to start your appreciation of mathematics.

It is natural to ask who first found Theorem 2. The question sounds simple, but the answer is not. The notation that we use today is only a recent development, and mathematicians who worked in ancient times and in the early middle ages used long complicated sentences to describe their results. Negative numbers and square roots were not allowed in the number systems of the early algebraists. So they could not possibly state the quadratic formula in the form we see today. Tne first steps toward a general method of solving a quadratic equation were taken by two Arab mathematicians, Musa al-Khowarizmi (who lived about 820 A.D.), and Abu Kamil (who lived about 850–930 A.D.).

We return to Equation (18), to see if it contains any additional information. It does. The $\pm$ sign indicates that a quadratic equation always (?) has two roots. But suppose that $b^2 - 4ac = 0$. Then the two roots are $-b/2a \pm 0$ and the two roots have coalesced into one root. What is worse: suppose that $b^2 - 4ac = a$ negative number, then the two roots are

$$r_1, r_2 = \frac{-b \pm \sqrt{\text{negative number}}}{2a}. \tag{20}$$

Thus a very nice theorem is within reach, namely:

Theorem 3. *Every quadratic equation has exactly two roots.*

We would like this theorem to be true. It is true, if we are willing to make two changes in our thinking. The first change is easy. Instead of counting the solutions we count the number of factors. For example $x^2 - 14x + 49 = 0$, is the same as $(x-7)(x-7) = 0$ so the root $x = 7$ is regarded as occurring twice. We call it a *repeated root* or a *multiple root of order two*. Thus, when counting roots we count them in accordance with their multiplicity. Therefore, the quadratic equation $x^2 - 14x + 49 = 0$ has two roots, $r_1 = 7$ and $r_2 = 7$ when counted in accordance with their multiplicity. This change in our thinking was very easy. The second change is much more difficult.

To make Theorem 3 true, we are forced to accept complex numbers into our system. Thus we must admit that numbers of the type given in Equation (20) are on an equal footing with our usual numbers such as 5, -9, $3/4$, $\sqrt{15}$, etc. Such acceptance was very difficult for many mathematicians, and only in the last 200 years were these numbers given their proper place. A rigorous presentation of the complex number system will be given in Chapter 25. Here, a brief sketch will be sufficient.

We adjoin to the real numbers a new number represented by the letter (symbol) i, with the special property that $ii = i^2 = -1$, so that $i = \sqrt{-1}$. The set of all complex numbers is the set of all numbers c that can be written in the form

$$c = a + bi \quad (\text{or } a + ib) \tag{21}$$

where a and b are real numbers. The number a is called *the real part of* c and b is called *the imaginary part of* c.

We *assume* that all the usual rules of algebra still hold. For example $\sqrt{-16} = \sqrt{16(-1)} = \sqrt{16}\,\sqrt{-1} = 4i$. Further $7i + 13i = 20i$.

Example 2. Solve the equation

$$x^2 + 5x + 11 = 0\,. \tag{22}$$

Solution. We do not immediately see the factors so we apply the quadratic formula (18). Here $a = 1$, $b = 5$, and $c = 11$, so Equation (18) gives

$$\begin{aligned} x &= \frac{-b \pm \sqrt{b^2 - 4ac}}{2a} = \frac{-5 \pm \sqrt{(5)^2 - 4(1)(11)}}{2(1)} = \frac{-5 \pm \sqrt{-19}}{2} \\ &= \frac{-5 \pm \sqrt{19(-1)}}{2} = \frac{-5 \pm \sqrt{19}\sqrt{-1}}{2} = \frac{-5}{2} \pm \frac{\sqrt{19}}{2} i\,. \end{aligned}$$

Consequently, the quadratic Equation (22) has two solutions. The real part of each solution is $-5/2$ and the imaginary parts of the solutions are $\pm\sqrt{19}/2$. ▶▶

It is obvious that we could NOT have guessed at the factors of $x^2 + 5x + 11 = 0$, without first solving the equation. Once we know the roots r_1 and r_2 theory tells us that the two factors are $x - r_1$ and $x - r_2$.

WARNING. We have broken our own rule "do not use one symbol with two different meanings in the same problem". First: a, b, and c denote the coefficients in the quadratic expression (12). Then we used the same letters to denote the complex number $c = a + bi$ in Equation (21). While this is certainly a bad practice, it can be accepted here because the context (form, use) tells you which meaning is intended. The letters that we used are the standard (usual) letters both for the quadratic formula and for complex numbers. Thus it would have been very difficult to use two different symbols for c.

Can we predict what kind of roots an equation has without finding the roots. The anwser is yes. Look at $b^2 - 4ac$, the quantity under the radical in the quadratic formula. If $b^2 - 4ac = 0$, then the two roots are the same. If $b^2 - 4ac$ is negative, then the two roots are complex numbers. Thus $b^2 - 4ac$ allows us to discriminate among the roots:

Definition 1 (Discriminant). The quantity $D = b^2 - 4ac$ is called the *discriminant* of the equation $ax^2 + bx + c = 0$.

Theorem 4. *Suppose that the coefficients of $ax^2 + bx + c = 0$ are all real numbers.*
If $D = b^2 - 4ac > 0$, then the roots of $ax^2 + bx + c = 0$ are real and distinct (different).
If $D = b^2 - 4ac = 0$, then the two roots of $ax^2 + bx + c = 0$ are both $-b/2a$ (repeated roots).
If $D = b^2 - 4ac < 0$, then the two roots of $ax^2 + bx + c = 0$ are both complex and different.

We have already given the proof of Theorem 4 in our earlier work.

Example 3. What can we say about the roots of the equation

$$13x^2 + 432x + 3589 = 0$$

without solving the equation?

Solution. In this equation $a = 13$, $b = 432$ and $c = 3589$. Hence $b^2 - 4ac = (432)^2 - 4(13)(3589) = 186624 - 186628 = -4 < 0$. Therefore the given equation has two complex roots. ►►

Suppose that some of the coefficients in the equation are complex numbers. The quadratic formula will still give the roots. (Isn't that a fantastic surprise?) If $b^2 - 4ac = 0$, the roots will still be reapeted, but the other two conclusions of Theorem 4 may be false.

Example 4. Solve the equation

$$x^2 - (2i + 3)x + 6i = 0\,. \tag{23}$$

Solution. It would take a person with very sharp eyes to see the factors. But we can always rely on the quadratic formula. Here, $a = 1$, $b = -2i - 3$, and $c = 6i$. Then

$$b^2 - 4ac = [-(2i + 3)]^2 - 4(1)6i = (2i)^2 + 12i + 9 - 24i = -4 + 9 - 12i = 5 - 12i\,.$$

Fortunately, this is a perfect square, $(3 - 2i)^2$, (see Problems 29 through 32 of the next excercise). Hence

$$\frac{-b \pm \sqrt{b^2 - 4ac}}{2a} = \frac{2i + 3 \pm \sqrt{5 - 12i}}{2} = \frac{2i + 3 \pm (3 - 2i)}{2}\,.$$

The $+$ sign gives $r_1 = 6/2 = 3$. The $-$ sign gives $r_2 = (2i + 2i)/2 = 2i$, so the two roots are 3 and $2i$. We should check our answers. Using $x = 2i$ in Equation (23) we find that

$$x^2 - (2i + 3)x + 6i = (2i)^2 - (2i + 3)(2i) + 6i = -4 + 4 - 6i + 6i = 0\,.$$

We leave it for the reader to check that the root $r_1 = 3$ also satisfies Equation (23). ►►

Inspired by the solution of the general quadratic equation, we should look for a similar formula that solves all cubic equations

$$ax^3 + bx^2 + cx + d = 0\,. \tag{24}$$

This formula was found by Tartaglia (1500–1557 A.D.) but it is far more complicated than the quadratic formula. The solution of the general cubic can be fund in many older algebra books, but has been dropped from most recent books on elementary algebra. The formula is now known as Cardan's formula after Hieronimo Cardano (1501–1576 A.D.) who stole the general solution from Tartaglia. We present this solution in Chapter 16, Volume 3.

As may be expected, the solution of the general fourth degree equation (the quartic) is still more difficult than the solution of the cubic. This complicated solution was first achieved by Ludovico Ferrari (1522–1565 A.D.).

The search was on for the solution of the general fifth degree equation (the quintic), but the search ended surprisingly when Paolo Ruffini (1765–1822 A.D.) proved the impossibility of such a formula. Somewhat later, but with greater rigor Niels Abel (1802–1829 A.D.) obtained the same result.

Exercise 2

(A)

In Problems 1 through 20 solve the given equation. Certainly you may use the quadratic formula. That is what it is for.

1. $x^2 + 4x + 2 = 0.$

2. $x^2 - 5x + 5 = 0.$

3. $x^2 - 10x + 10 = 0.$

4. $x^2 + 10x - 1 = 0.$

5. $x = 3 + 3/x.$

6. $3x^2 + 2x = 1.$

7. $2x^2 = 8x + 1.$

8. $x^2 + 6x + 6 = 0.$

9. $2x^2 - 7x + 6 = 0.$

10. $2x^2 + 8x + 3 = 0.$

11. $2x^2 - 10x + 7 = 0.$

12. $6x^2 - 7x + 2 = 0.$

13. $x^2 + 4x + 11 = 0.$

14. $3x^2 + 6x + 1 = 0.$

15. $5x^2 - 2x + 3 = 0.$

16. $x^2 + 2x + 24 = 0.$

17. $5x^2 - 6x + 7 = 0.$

18. $x^2 + 5x + 24 = 0.$

19. $x^3 - 1 = 0$. Hint: $x - 1$ is a factor.

***20.** $x^3 - 2x + 1 = 0.$

***21.** Find all the roots of $x^2 + 2x + \dfrac{56}{x^2 + 2x} = 15$. Hint: First set $u = x^2 + 2x$. Then solve $u + 56/u = 15$.

***22.** Find all the roots of $x^2 - 3x - \dfrac{8}{x^2 - 3x} = 2.$

23. Give an example of a polynomial equation which has a root of multiplicity 3.

24. Give a polynomial equation in which -2 is a root of multiplicity 4.

If you review the proof of the quadratic formula, you will find that all of the arguments are valid for complex numbers. Therefore the formula should be usable on quadratic equations with complex coefficients. In Problems 25 through 28 use the quadratic formula to solve the equation and then check that your solutions do satisfy the given equation.

***25.** $x^2 - 2(1+i)x + 2i - 1 = 0.$

26. $x^2 - 4ix + 5 = 0.$

27. $x^2 - (3+2i)x + 1 + 3i = 0.$

28. $ix^2 + 5x + 6i = 0.$

29. In Example 4 we stated that $(5 - 12i)^{1/2} = \pm(3 - 2i)$. Check that this is true by squaring $(3 - 2i)$.

***30.** Problem 29 was easy because we were given the answer, $3 - 2i$. If we do not know the square root, we may try to find it by setting $(A + iB)^2 = C + iD$, where C and D are known and we are trying to find the unknowns A and B to obtain the square root of $C + iD$. Find the pair of equations that must be satisfied, to obtain the square root of $C + iD$.

31. Use the equations from Problem 30 to find one of the square roots of $80 + 18i$. The other square root is the negative of the one you found.

***32.** Solve the equation $x^2 + (3 - i)x - 18 - 6i = 0$. Hint: Use the result of Problem 31.

In Problems 33 through 38 find the two square roots of the given complex number.

33. $-3 + 4i.$

34. $-24 - 10i.$

35. $15 + 8i.$

36. $-21 + 20i.$

37. $21 + 20i.$

38. $65 - 72i.$

How to find the nth roots of an arbitrary complex number will be covered in Theorems 13 and 14 of Chapter 25.

(B)

As you may have noticed most problems are composed so that the solutions will be "nice" numbers (integers, or rational numbers with reasonable denominatiors). In any problems that arises in a *natural* way, the solutions will not always be so "nice". In the following problems use the quadratic formula and a calculator to give approximate roots to the nearest hundrenth. In problem sets labelled (A) you are supposed to work the problem by hand (without calculator or computer). In problem sets labeled (B) you are expected to use these wonderful new tools.

39. $1.23x^2 + 3.45x - 6.78 = 0.$

40. $4.68x^2 - 1.35x - 5.79 = 0.$

41. $1776x^2 - 1812x - 1984 = 0.$

42. $40.40x^2 + 33.33x - 10.11 = 0.$

3. Other Types of Equations

When trying to solve some equations, the work may lead to an equivalent quadratic equation, although this was not expected at the start.

Example 1. Solve the equation

$$\frac{3}{x-2}+\frac{4}{x-3}-\frac{3}{x-4}=0\,. \tag{25}$$

Solution. We multipy both sides of (25) by the L.C.D. which is $(x-2)(x-3)(x-4)$ and obtain

$$\begin{aligned}0&=3(x-3)(x-4)+4(x-2)(x-4)-3(x-2)(x-3)\\&=3(x^2-7x+12)+4(x^2-6x+8)-3(x^2-5x+6)\\&=(3+4-3)x^2+(-21-24+15)x+(36+32-18)=4x^2-30x+50\,.\end{aligned}$$

Thus we have $2(2x^2-15x+25)=2(2x-5)(x-5)=0$. Thus $x=5/2,5$.

To check the solution $x=5/2=2.5$, we substitute this in Equation (25) and find that

$$\begin{aligned}\frac{3}{2.5-2}+\frac{4}{2.5-3}-\frac{3}{2.5-4}&=\frac{3}{1/2}+\frac{4}{-1/2}-\frac{3}{-3/2}\\&=\;6\;-\;8\;+\;2\;=0\,.\end{aligned}$$

We leave it for the student to check the solution $x=5$. ▶▶

Example 2. Solve the equation

$$2\sqrt{x+6}-x=3\,. \tag{26}$$

Solution. We transpose x to the right side and then square both sides. This gives $(2\sqrt{x+6}\,)^2=(x+3)^2$, or

$$4(x+6)=(x+3)^2=x^2+6x+9\,.$$

We put all the terms on the right side and find that

$$0=x^2+6x-4x+9-24=x^2+2x-15=(x+5)(x-3)\,.$$

Hence, $x=3$ and $x=-5$ are possible solutions.

To check the solution $x=3$, we substitute this value in the right side of (26). This gives

$$2\sqrt{x+6}-x=2\sqrt{3+6}-3=2(3)-3=3\,,$$

and so $x=3$ is a solution. But when we set $x=-5$ in the right side of (26) we obtain

$$2\sqrt{x+6}-x=2\sqrt{-5+6}-(-5)=2+5=7\,,$$

and not 3 as we should. Thus $x = -5$ is not a solution of (26). How did this error(?) creep in? The answer is simple. When we square both sides of an equation $F = G$, to obtain $F^2 = G^2$, the new equation is not equivalent to the original one because F might be negative and G might be positive. Thus, $-5 \neq 5$, but their squares are equal. The operation of squaring both sides of an equation may introduce new roots that are not roots of the original equation. Such numbers are called *extraneous roots.* In this problem $x = -5$ is an extraneous root. Thus Equation (26) has only one solution, $x = 3$. ▶▶

Example 3. Solve the equation

$$\sqrt{2x+4} + \sqrt{2x-3} = \sqrt{6x+13}\,. \tag{27}$$

Solution. It seems as though squaring will be needed. In Example 2, it was helpful to transpose one term before squaring. Here, such a move will not help, so we square both sides of (27), keeping in mind that all the roots of the original equation are still roots of the new equation, but that some extra roots may show up. After squaring we get

$$2x + 4 + 2\sqrt{2x+4}\,\sqrt{2x-3} + 2x - 3 = 6x + 13\,,$$

or, after moving some terms to the right side

$$\begin{aligned} 2\sqrt{2x+4}\,\sqrt{2x-3} &= 6x + 13 - (2x+4) - (2x-3) \\ &= 2x + 12 = 2(x+6)\,. \end{aligned}$$

We divide by 2, and then square both sides once more. This gives

$$(2x+4)(2x-3) = (x+6)^2 = x^2 + 12x + 36\,,$$

or

$$\begin{aligned} 4x^2 + 8x - 6x - 12 &= x^2 + 12x + 36\,, \\ 3x^2 - 10x - 48 &= 0\,, \\ (3x+8)(x-6) &= 0\,. \end{aligned}$$

Consequently, our work leads to the solutions $x = 6$, and $x = -8/3$.

Check: When $x = 6$, Equation (27) becomes

$$\sqrt{2(6)+4} + \sqrt{2(6)-3} = \sqrt{6(6)+13}\,.$$

or $\sqrt{16} + \sqrt{9} = \sqrt{49}$, or $4 + 3 = 7$, which is true. When $x = -8/3$ the second term in (27) is the square root of a negative number. Consequently, no further work is needed. The only solution to equation (27) is $x = 6$. ▶▶

Exercise 3

(A)

In Problems 1 through 31 solve the given equation.

1. $\dfrac{5}{x+2}=\dfrac{3}{x+1}+\dfrac{2}{x+3}.$

2. $\dfrac{1}{x-5}-\dfrac{4}{x-2}+\dfrac{3}{x+1}=0.$

3. $\dfrac{2}{x-9}=\dfrac{3}{x-7}+\dfrac{2}{x-3}.$

4. $\dfrac{1}{x-3}+\dfrac{1}{x+1}-\dfrac{12}{x+6}=0.$

5. $\dfrac{1}{x-7}+\dfrac{2}{x-1}-\dfrac{2}{x+2}=0.$

6. $\dfrac{1}{2x-1}+\dfrac{2}{x+1}-\dfrac{3}{4x-5}=0.$

7. $\dfrac{1}{x-1}+\dfrac{1}{x-2}+\dfrac{1}{x-3}=0.$

8. $\dfrac{2}{x+2}+\dfrac{2}{x+3}-\dfrac{7}{x+5}=0.$

9. $\dfrac{1}{x+1}+\dfrac{2}{x+3}-\dfrac{5}{x+7}=0.$

10. $\dfrac{1}{x+1}+\dfrac{1}{x+2}+\dfrac{1}{x+3}=0.$

11. $\sqrt{x+6}+2x=9.$

12. $\sqrt{x+7}+5=x.$

13. $\sqrt{4x-2}+5=2x.$

14. $\sqrt{x+23}-\sqrt{x+16}=1.$

15. $\sqrt{x+13}-\sqrt{x+5}=2.$

16. $\sqrt{5x+21}-2=\sqrt{x+13}.$

17. $\sqrt{x+2}+\sqrt{x+16}=7.$

18. $\sqrt{x+19}+\sqrt{x+1}=6.$

19. $\sqrt{2x+3}+\sqrt{3x+4}=\sqrt{5x+9}.$

20. $\sqrt{7x+14}-2=x.$

21. $\sqrt{2x-7}-1=\sqrt{x-4}.$

22. $\sqrt{x-1}+\sqrt{3-x}=\sqrt{x+2}.$

23. $\sqrt{6-x}=\sqrt{2x+5}-\sqrt{x-1}.$

24. $\sqrt{2x+3}=\sqrt{5x+8}-\sqrt{3x+5}.$

25. $3\sqrt{x}=\dfrac{20}{\sqrt{x}}-11.$

26. $6\sqrt{x}=\dfrac{5}{\sqrt{x}}-13.$

***27.** $\sqrt{\dfrac{x}{1-x}}+\sqrt{\dfrac{1-x}{x}}=\dfrac{5}{2}.$ Hint: First solve $u+\dfrac{1}{u}=\dfrac{5}{2}.$

***28.** $\sqrt{\dfrac{2x}{1-3x}}+\sqrt{\dfrac{1-3x}{2x}}=\dfrac{13}{6}.$ Hint: First solve $u+\dfrac{1}{u}=\dfrac{13}{6}.$

***29.** $\sqrt{2x^2-9x+4}-\sqrt{2x^2-7x+1}=1.$

***30.** $\sqrt{2x^2+5x-2}-\sqrt{2x^2+5x-9}=1.$

***31.** $\dfrac{x+\sqrt{x^2-1}}{x-\sqrt{x^2-1}}-\dfrac{x-\sqrt{x^2-1}}{x+\sqrt{x^2-1}}=8x\sqrt{x^2-3x+2}.$

32. $\sqrt{7+\sqrt{x}}=\sqrt{x}+1.$ **34.** $\sqrt{13+4\sqrt{x}}=\sqrt{x}+2.$ **33.** $\sqrt{9-2\sqrt{x}}=\sqrt{x}-1.$

4. Applications of Quadratic Equations

Some natural problems lead to quadratic equations. In many cases, the problems are not natural, but are created merely to test the ability of the student to transform the words into equations. Perhaps you have already noticed such problems in your earlier work. Whether the problems are natural or not, they provide good exercise for the brain and may actually be fun.

Example 1. We want to make a box by cutting a 4 inch square from each corner of a flat square piece of tin, and turning up the edges (see Figure 1). Find the size of the orginal piece of tin, if the finished box has no top and we want the box to have a volume of 100 cubic inches.

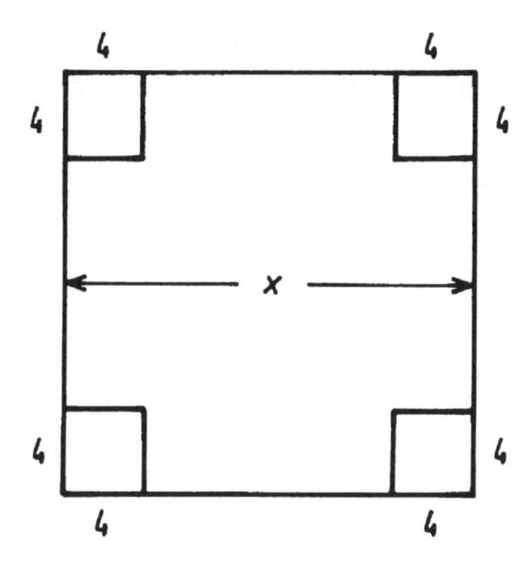

Figure 1.

Solution. Let $x =$ the length of the side of the large square. After cutting the small squares from the corners, the length of one edge of the base will be $x - 2(4) = x - 8$. Therefore the volume will be (height)(Area of base) or

$$\begin{aligned} V &= 4(x-8)(x-8) = 100\,, \\ V/4 &= x^2 - 16x + 64 = 100/4 = 25\,. \end{aligned} \tag{28}$$

Thus,

$$x^2 - 16x + 39 = 0\,,$$

$$(x-13)(x-3) = 0\,.$$

Therefore $x = 3$ or $x = 13$. But if we go back to the source of the problem, the answer $x = 3$ is ridiculous because one cannot cut a 4 inch square from the corners of a piece that is only 3 inches by 3 inches. Hence the original piece of tin should be 13 inches by 13 inches.

Check: After removing the little 4 inch squares, the length of one side of the base is $13 - 2(4) = 5$. Then the volume of the finished box is $5(5)(4) = 100$ in^3. ▶▶

Remark. One might argue that quadratic equations are not necessary here because one can almost guess the answer. We merely check original sizes for $x = 9, 10, 11, 12, 13$, etc. until we find the one that works. But suppose that we want the final volume to be 112 cubic inches? Then Equation (28) becomes

$$V/4 = x^2 - 16x + 64 = 112/4 = 28\,, \quad \text{or} \quad x^2 - 16x + 36 = 0\,,$$

and eventually

$$x = 8 \pm 2\sqrt{7}\,,$$

where only the $+$ sign gives a meaningful answer. Observe that when the answer is not an integer, it would be impossible to guess the result. In such a situation we MUST use algebra if we want the correct result.

Example 2. The sum of the squares of two adjacent odd integers is 290. Find the integers.

Solution. Let x and y be the two numbers with y the larger of the two. Since they are adjacent odd integers, $y = x + 2$. The sum of the squares gives the equation

$$\begin{aligned} 290 &= x^2 + y^2 = x^2 + (x+2)^2 = x^2 + (x^2 + 4x + 4) \\ &= 2x^2 + 4x + 4\,, \end{aligned} \tag{29}$$

or

$$145 = x^2 + 4x + 2\,,$$

or

$$0 = x^2 + 2x - 143 = (x - 11)(x + 13)\,.$$

So $x = 11$, or -13. But $y = x + 2$ so the pairs are $(11, 13)$ and $(-13, -11)$. ▶▶

One might have guessed the solution $(11, 13)$ but the student who guesses might have missed the second solution $(-13, -11)$. Further, the problem specified that the numbers be integers. This makes it easy to guess the solution. Suppose, however, that the condition that the numbers be integers is removed and the sum of the squares is changed slightly. Then it becomes absolutely necessary to use algebra. This is illustrated in

Example 3. The sum of the squares of two numbers is 280. If the numbers differ by 2, find the numbers.

Solution. As in Example 2 we let x and y be the two numbers with $y = x + 2$. As before, we arrive at Equation (29) but with 290 replaced by 280. Then, after dividing by 2 we have

$$140 = x^2 + 2x + 2$$

or

$$0 = x^2 + 2x - 138\,.$$

The quadratic formula gives

$$x = \frac{-2 \pm \sqrt{4 + 4(138)}}{2} = \frac{-2 \pm 2\sqrt{1 + 138}}{2}$$
$$= -1 \pm \sqrt{1 + 138} = -1 \pm \sqrt{139} \quad \text{(after canceling 2)}.$$

As in Example 2 we have two solutions (x, y),

$$(-1 + \sqrt{139}, 1 + \sqrt{139}) \quad \text{and} \quad (-1 - \sqrt{139}, 1 - \sqrt{139}). \qquad ▶▶$$

Clearly, no amount of guessing would have found either one of these solutions.

Example 4. It is well known that a triangle with sides 6, 8 and 10 feet long is a right triangle with the hypotenuse being the longest side. How many feet must be added to each of the sides to give a right triangle with a 20 foot hypotenuse.

Solution. By the wording of the problem we understand that the same amount must be added to each side. Let d be the added amount. Then to have a right triangle we must satisfy the condition

$$(6 + d)^2 + (8 + d)^2 = (20)^2 = 400. \tag{30}$$

Therefore

$$36 + 12d + d^2 + 64 + 16d + d^2 = 400,$$
$$2d^2 + (12 + 16)d + 36 + 64 - 400 = 0,$$
$$2d^2 + 28d - 300 = 0,$$
$$d^2 + 14d - 150 = 0. \tag{31}$$

The factors are not obvious, so we apply the quadratic formula to (31) and find that

$$d = \frac{-14 \pm \sqrt{(14)(14) - 4(1)(-150)}}{2}$$
$$= \frac{-2(7) \pm \sqrt{(4)(7)(7) + 4(150)}}{2}$$
$$= -7 \pm \sqrt{49 + 150},$$
$$d = -7 + \sqrt{199}. \qquad ▶▶$$

Notice that in the last step we discarded the — sign because in this problem triangles with lengths that are negative make no sense.

The amount to be added is $-7 + \sqrt{199} =$ approx. 7.107.

To check the solution, we observe that the lengths of the two sides are $x = 6 + d = 6 - 7 + \sqrt{199}$ and $y = 8 + d = 8 + -7 + \sqrt{199}$. Thus

$$x^2 + y^2 = (-1 + \sqrt{199})^2 + (1 + \sqrt{199})^2$$
$$= [1 - 2\sqrt{199} + 199] + [1 + 2\sqrt{199} + 199] = 1 + 199 + 1 + 199$$
$$= 400. \qquad ▶▶$$

Example 5. Find k so that the equation

$$x^2 - 2(k+3)x + k^2 = 0 \tag{32}$$

has repeated roots. What is the repeated root?

Solution. The quadratic equation will have repeated roots if and only if the discriminant $D = b^2 - 4ac = 0$. Here we have $b = -2(k+3)$ and $c = k^2$, so

$$\begin{aligned} D = b^2 - 4ac &= [-2(k+3)]^2 - 4(1)k^2 \\ &= 4(k^2 + 6k + 9) - 4k^2 \\ &= 4(6k+9) = 0\,. \end{aligned}$$

Therefore $k = -9/6 = -3/2$. With this k, Equation (32) becomes

$$\begin{aligned} x^2 - 2(-3/2+3)x + (-3/2)^2 = 0\,, \\ x^2 - 3x + 9/4 = (x - 3/2)^2 = 0\,. \end{aligned}$$

Thus $k = -3/2$, and the repeated root is $r = 3/2$. ►►

Example 6. Find two numbers that differ by 6 and such that the sum of their squares is 28.

Solution. Let x and y be the two numbers, and suppose that without loss of generality we let x be the larger one. Then we have the two equations

$$x = y + 6\,, \tag{33}$$

$$x^2 + y^2 = 28\,. \tag{34}$$

If we substitute the expression for x from Equation (33) into Equation (34) we have

$$\begin{aligned} (y+6)^2 + y^2 = 28\,, \\ y^2 + 12y + 36 + y^2 - 28 = 0\,, \\ 2y^2 + 12y + 8 = 0\,, \end{aligned}$$

or, if we divide by 2

$$y^2 + 6y + 4 = 0\,.$$

We apply the quadratic formula to this equation in y (the name of the variable is not important).

$$\begin{aligned} y &= \frac{-6 \pm \sqrt{36-16}}{2} \\ &= \frac{-6 \pm \sqrt{4(9-4)}}{2} \\ &= -3 \pm \sqrt{5}\,. \end{aligned}$$

Thus there are two sets of answers (I) $y = -3 - \sqrt{5}$, $x = 3 - \sqrt{5}$, and (II) $y = -3 + \sqrt{5}$, $x = 3 + \sqrt{5}$. The reader is left to show that x is 6 larger than y in both (I) and (II), and that in each case the sum of the squares is 28. Observe that in each solution one of the numbers is negative and one is positive. Further in (II) each number is the negative of one of the numbers in (I). ►►

Exercise 4

(A)

1. Find the dimensions of a rectangle if its area is 88 square feet and the length is 3 feet more than the width.

2. Do Problem 1 if the area is 108 square feet.

3. Do Problem 1 if the area is 100 square feet and the length is 4 feet more than the width.

4. Do Problem 1 if the area is 90 square feet and the width is 6 feet less than the length.

5. If a border x feet wide is added all around a 9 ft by 12 ft rug, the area of the rug is increased by 72 square feet. Find the width of the border.

6. Do Problem 5 if the rug is 7 ft by 12 ft and the area is increased by 66 square feet.

7. Find two numbers that differ by 6, and for which the sum of their squares is 22.

8. Do Problem 7, if the difference of the two numbers is 5 and the sum of their squares is 91.

9. Find two numbers whose sum is 8 and whose product is 13.

10. Do Problem 9 if the sum is 9 and the product is 14.

11. Do Problem 9 if the sum is 10 and the product is 15.

12. Do Problem 9 if the sum is 30 and the product is 198.

In Problems 13 through 18 find the value of k that forces the given equation to have a repeated root. Find the repeated root.

13. $x^2 - 2(k+3)x + k^2 + 3 = 0$.

14. $x^2 - 2(k+3)x + k^2 - 9 = 0$.

15. $x^2 - (2k-1)x + k^2 = 0$.

16. $x^2 + (k-2)x + k^2 = 0$.

17. $x^2 - (k+3)x + 4k = 0$.

***18.** $kx^2 + (k+2)x + k + 1 = 0$.

***19.** Suppose that in the equation $x^2+2(k+b)x+k^2=0$, (with b fixed) we select k so that the equation has repeated roots. Prove that for every b it is true that k is the same as the repeated root.

In Problems 20 through 23 we are given the difference of two numbers x and y, and also the difference of their cubes. Find the two numbers. Recall that we proved earlier that

$$(A+B)^3 = A^3 + 3A^2B + 3AB^2 + B^3\,.$$

20. $x-y=1$, and $x^3-y^3=10$.

21. $x-y=2$, and $x^3-y^3=98$.

22. $x-y=2$, and $x^3-y^3=32$.

***23.** $x-y=-1,\quad x^3-y^3=8$.

***24.** Check that the solutions (I) to Problem 23 satisfy the given conditions, $x^3-y^3=8$.

Chapter 8

Graphing Equations

1. The Rectangular Coordinate System

Before the sixteenth century algebra and geometry were regarded as separate subjects. It was René Descartes (1596–1650) who first noticed that these two subjects could be united and that each subject could contribute to the development of the other. This union, which we now call Analytic Geometry, has been fruitful far beyond the wildest dreams of Descartes. The unifying element is the rectangular coordinate system. Probably the reader is already familiar with this system. Completeness requires that we present it (once more) to the reader.

The number line (alias the x-axis, alias the horizontal axis, alias the real axis) has been explained previously in Chapter 2, Section 2, but it is sufficiently important that we review it here. We draw a line, horizontally for convenience, and select a fixed point O to represent the number 0 (zero). We select a unit of measurement and a positive direction on the line, usually to the right, as indicated by the arrow (see Figure 1). Once these items are settled, each point P on the number line is associated with a unique number, called the coordinate of the point. If $d \geq 0$ is the distance of P from the origin, then the coordinate of P is d, if P lies to the right of the origin. If P lies to the left of the origin, then the coordinate of P is $-d$. The coordinate of the origin O is 0 (zero). Some points together with their coordinates are shown in Figure 1.

We now introduce the concept of the *directed distance* from P to Q, which is really the distance between P and Q with a negative sign attached when going the "wrong way".

Definition 1 (Directed distance). Let $d > 0$ be the distance between two points P and Q on a directed line (a line with a positive direction). If P precedes Q on the line, then the directed distance PQ is d. If Q precedes P on the line, then the directed distance PQ is $-d$. If the two points coincide, then the directed distance $PQ = 0$.

We use the term "precedes" because the number line might not be horizontal. The point P precedes Q if PQ has the same direction as the arrow on the line. If the line is horizontal

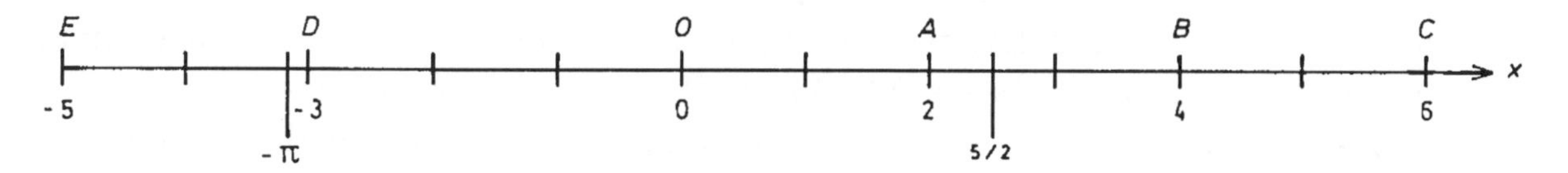

Figure 1.

then P precedes Q if Q is to the right of P. For example the points in Figure 1 give the following directed distances:

$$AB = 2\,, \quad BA = -2\,, \quad DA = 5\,, \quad AD = -5\,, \quad EC = 11\,, \quad CE = -11\,,$$
$$OB = 4\,, \quad BO = -4\,, \quad OC = 6\,, \quad CO = -6\,, \quad EO = 5\,, \quad OE = -5\,.$$

Theorem 1. *For any three points P, Q, and R on a number line, the directed distances satisfy*

$$\boxed{PQ + QR = PR\,.} \tag{1}$$

This equation is easy to memorize, because we can just suppress the middle letter Q on the left side to obtain the right side.

For example, referring to the points in Figure 1,

$DA + AC = DC\,,$ Theorem 1 states that $5 + 4 = 9\,.$
$AD + DC = AC\,,$ Theorem 1 states that $-5 + 9 = 4\,.$
$CD + DA = CA\,,$ Theorem 1 states that $-9 + 5 = -4\,.$
$BD + DA = BA\,,$ Theorem 1 states that $-7 + 5 = -2\,.$
$CE + EA = CA\,,$ Theorem 1 states that $-11 + 7 = -4\,.$

All of these statements are correct. Indeed the right side is always a consequence of Theorem 1. These five lines are a check to see if Theorem 1 is true. The reader should compose more checks of this type for his own amusement and understanding. One must realize that no matter how many examples one checks THIS WILL NEVER BE A PROOF OF THEOREM 1. However, if just one example gives a false result this is enough to show that the theorem that you are checking is not true.

To prove Theorem 1, we first observe that the letters P, Q, and R can be placed on a number line in only six different ways as far as order is concerned. With each of these ways we indicate the various distances by letters x, y, and z, and then persuade ourselves (or whoever is listening) that in each of the six cases, Equation (1) is correct.

We have already agreed that the directed distance from P to Q is indicated by the symbols PQ. It is convenient to use the symbols $|PQ|$ for the distance, a number that is always positive (or zero if P and Q coincide).

With these elaborate preparations we are now ready for the rectangular coordinate system. The rectangular coordinate system for a plane is obtained by introducing two number lines meeting at right angles at a point O called the *origin*. It is customary to make one of these lines horizontal with the positive direction to the right and the second line vertical with the positive direction upward (see Figure 2). The horizontal line is called the *x-axis* (or *horizontal axis*) and the vertical line is called the *y-axis* (or the *vertical axis*). These two axes divide the plane into four quadrants labeled Q.I, Q.II, Q.III, and Q.IV (see Figure 2).

Once a rectangular coordinate system has been chosen, any point in the plane can be located with respect to it. Suppose that P is a point in the plane. Let QP be the line segment from P perpendicular to the x-axis at the point Q and let RP be the line segment

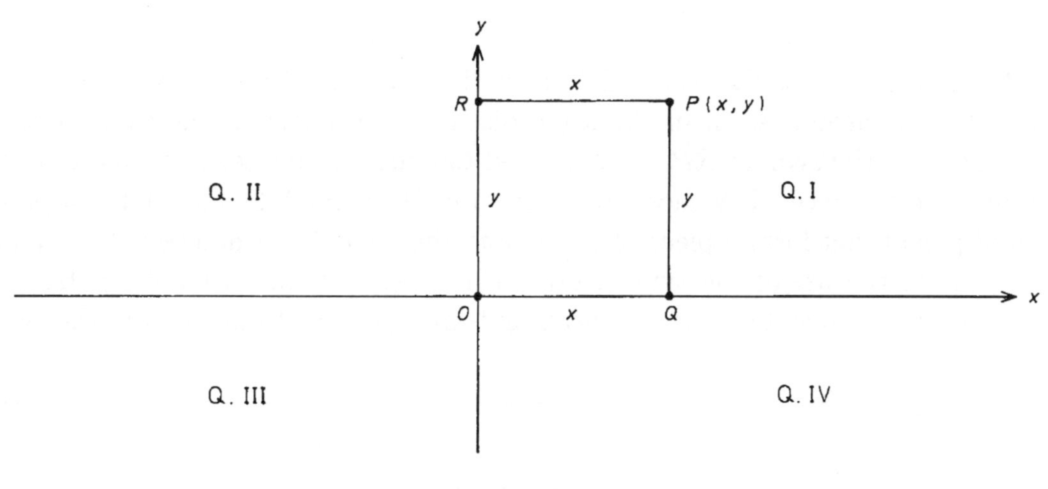

Figure 2.

from P perpendicular to the y-axis at R. Then the directed distance QP is called the y-coordinate of P and the directed distance RP is called the x-coordinate of P (see Figure 2). Please observe that as directed distances $QP = OR$ and $RP = OQ$, so that OR and OQ could also be used to define the y- and x-coordinates of a point. It is customary to indicate the coordinates of a point by enclosing them in parentheses thus: (x, y). A number of points together with their coordinates are shown in Figure 3.

Of course, this procedure can be reversed. Given any pair of real numbers x, and y, there is a unique point in the plane whose coordinates are (x, y). We summarize these remarks in.

Theorem 2. *With a given rectangular coordinate system each point in the plane has a uniquely determined pair of coordinates (x, y). Conversely, each pair of numbers x, y determines a unique point that has (x, y) for its coordinates.*

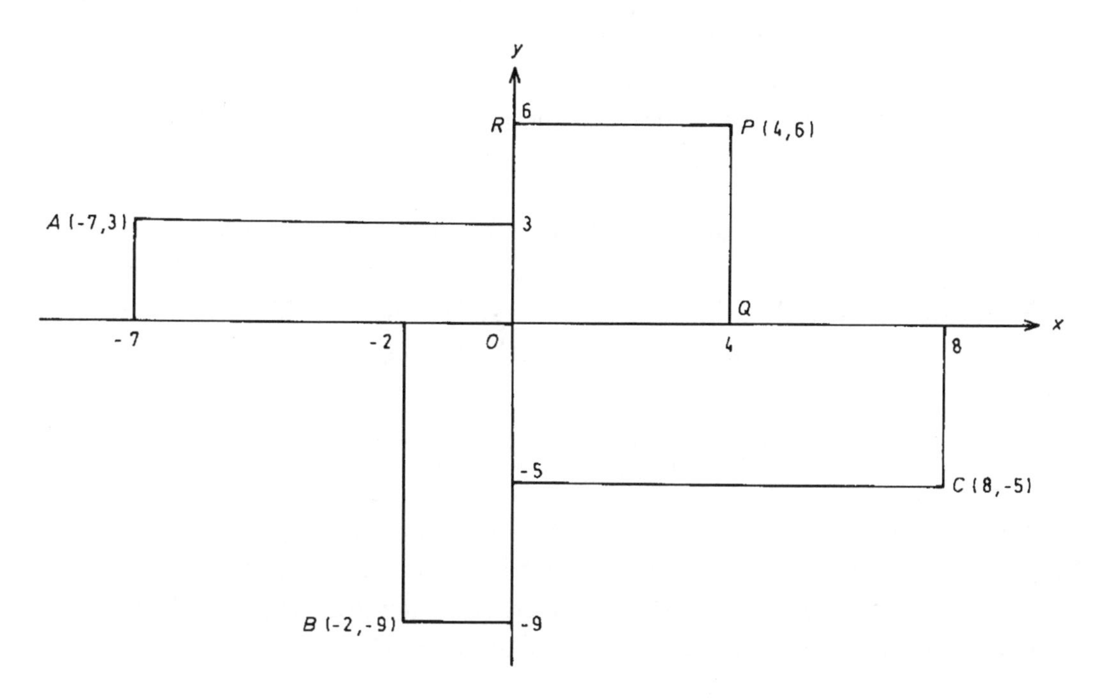

Figure 3.

AN APOLOGY. Perhaps the reader did not notice it at the time but we used one symbol with two different meanings, all in the same discussion, (in fact in the same paragraph). We are referring to the symbol RP which is used to denote a line segment and also used to denote a directed distance. Now these two items are DIFFERENT. The first is a particular collection of points that form a piece of a line, and the second is a number. This same type of error occurs in the use of the symbol QP (a line segment and a number). However no great harm has been done, because the meaning is always clear from the way the symbol is used.

A second apology. Some mathematicians may disagree with the title Theorem used in Theorem 2. Some would call it a Definition, the definition of the plane. Still other mathematicians would call it an Axiom, an axiom about the nature of the plane. We do not wish to argue about this item. We have more important things to do.

Our next project is the proof of a formula for the distance between two points $P(x_1, y_1)$ and $Q(x_2, y_2)$ in the plane. (Recall that the use and convenience of subscripts and superscripts has already been discussed in Chapter 4, Section 4.)

Theorem 3. *Let $P_1(x_1, y_1)$ and $P_2(x_2, y_2)$ be any two points in the plane. Then the distance between these two points is given by*

$$d = |P_1P_2| = \sqrt{(x_2 - x_1)^2 + (y_2 - y_1)^2}\,. \tag{2}$$

Note that in the case of a slanted line segment, P_1P_2 is NOT a directed distance. However we may still use the notation $|P_1P_2|$ for the distance between the points (or the distance from the first point to the second point).

Proof. As indicated in Figure 4 we make the line segment joining the points P_1 and P_2 into the hypotenuse of a right triangle, by drawing suitable lines. Suppose that the horizontal line through P_1 and the vertical line through P_2 meet at Q. Then the sides of the right triangle P_1QP_2 have lengths $a = |P_1Q|$ and $b = |QP_2|$. By the Pythagorean Theorem, the hypotenuse d is given by

$$d^2 = a^2 + b^2\,, \quad \text{or} \quad d = \sqrt{a^2 + b^2}\,. \tag{3}$$

All that remains is to find convenient expressions for a and b.

On the horizontal line through P_1, the y-coordinate is the constant y_1. On the vertical line through P_2, the x-coordinate is the constant x_2. Therefore the coordinates of Q, the intersection point of these two lines, are (x_2, y_1). Thus, $a = |P_1Q| = |EF|$. Now, by Theorem 1

$$P_1Q = EF = EO + OF = -x_1 + x_2\,, \tag{4}$$

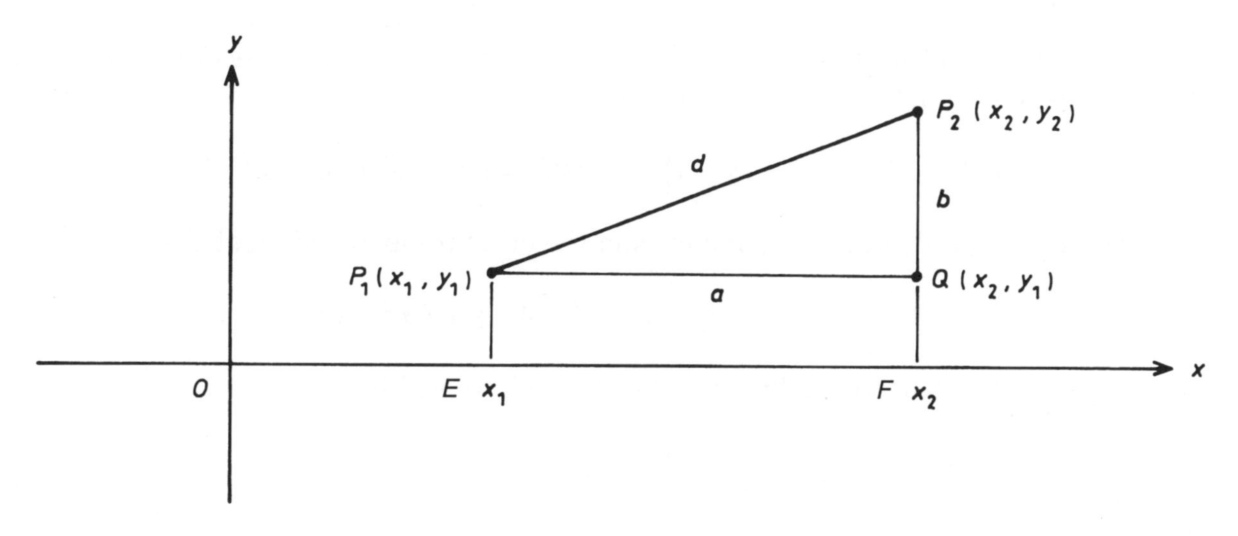

Figure 4.

by the definition of the x-coordinate on the horizontal number line. Consequently,

$$a^2 = |P_1Q|^2 = |EF|^2 = (-x_1 + x_2)^2 = (x_2 - x_1)^2 . \tag{5}$$

A similar argument will show that $b^2 = (y_2 - y_1)^2$. When these results are used in (3) we obtain equation (2) and the theorem is proved. ∎

Wait a minute! In presenting this proof, the argument was based on the picture in Figure 4. In that picture, both points are shown in the first quadrant, with P_2 to the right and above P_1. Selecting the points in these special positions simplifies the argument and makes it easier for the student to follow. However, the argument that has just been presented is valid, because a careful examination will show that each statement in the argument is true no matter where the two points P_1 and P_2 are located in the plane.

Example 1. Find the distance between the two points $(-3, 7)$ and $(2, -5)$.

Solution. We may select either point to be P_1 and then the other point is automatically P_2. If we select $(-3, 7)$ to be P_1, then P_2 is $(2, -5)$ and Equation (2) gives

$$d^2 = (2 - (-3))^2 + (-5 - 7)^2 = 5^2 + (-12)^2 = 25 + 144 = 169 ,$$

so $d = \sqrt{169} = 13$. ▶▶

We were lucky that 169 turned out to be a perfect square, the square of 13. We must not always expect such good fortune.

The reader is urged to make a careful drawing (to some convenient scale) showing the points in this example. He/She should then check the distance formula by direct measurement of P_1P_2 on the drawing.

The reader should also work this same example using $(2, -5)$ as P_1 and $(-3, 7)$ as P_2 and explain why one obtains 13, the same answer.

Example 2. If the point $P(x, y)$ is equidistant from the two points $A(4, 1)$ and $B(2, 3)$, find a simple equation that the coordinates of P must satisfy.

Solution. The problem states that $|PA| = |PB|$. If we square both sides of this equation and use Equation (2) of Theorem 3, we obtain

$$|PA|^2 = (x-4)^2 + (y-1)^2 = |PB|^2 = (x-2)^2 + (y-3)^2 .$$

If we drop the letters A, B, and P, and expand the other terms we will find that

$$x^2 - 8x + 16 + y^2 - 2y + 1 = x^2 - 4x + 4 + y^2 - 6y + 9 ,$$

or on combining terms and canceling where permitted, we will have

$$0 = 4x - 4y - 4 .$$

Thus

$$y = x - 1 . \tag{6}$$

All of the operations used in going from the condition $|PA| = |PB|$ to Equation (6) are reversible. Thus the coordinates of every point that satisfies the condition $|PA| = |PB|$ must satisfy Equation (6) and conversely if the coordinates satisfy Equation (6) then $|PA| = |PB|$. ▶▶

Equation (6) is the solution to the problem posed. We will see later that this is the equation of a straight line. In this case a basic theorem from geometry tells us that the collection of all the points whose coordinates satisfy Equation (6) is a line that is the perpendicular bisector of the line segment joining points A and B.

The rectangular coordinate system is frequently called the Cartesian coordinate system in honor of its inventor René Descartes (1596–1650). A brief and highly entertaining account of the life of this genius can be found in "Men of Mathematics" by E. T. Bell, Simon and Schuster 1937.

Exercise 1

(A)

1. Three points A, B, and C lie on the x-axis. Make a list showing all possible ways that they can be placed with respect to their order on the axis. For example, the order A, B, C means that as we proceed on the x-axis from left to right we meet A first, then B, and then C.

2. Suppose that A, B, and C lie on a directed line with $AB = 5$, and $BC = 8$. Write all possible equations of the form of Equation (1) for these three points, and use the numbers given for AB and BC to check that all of these forms give numerically correct results.

3. What is the formula for d^2 when P_1 is the origin. In this case the distance d is often denoted by r.

In Problems 4 through 7, find the missing coordinate of P when r is the distance of P from the origin.

4. $r = 5$, $x = 4$, P is in Q.IV (P is in the fourth quadrant).

5. $r = 13$, $y = -12$, P is in Q.III.

6. $r = \sqrt{29}$, $x = -2$, P is in Q.II.

7. $r = 2\sqrt{7}$, $y = 4$, P is in Q.I.

In Problems 8 through 16, describe the figure formed by the set of all points whose coordinates satisfy the given equation. (Later in this chapter we will make a systematic study of this type problem.)

8. $y = 6$.

9. $x = -3$.

10. $y = x$.

11. $y = -x$.

12. $y = x + 1$.

13. $x^2 + y^2 = 49$.

***14.** $x^2 = y^2$.

***15.** $x^3 = y^3$.

***16.** $x^4 = y^4$.

In Problems 17 through 20 find $|AB|$ for the given pair of points.

17. $A(-9, -1), B(3, 4)$.

18. $A(-2, 9), B(1, 5)$.

19. $A(7, 1), B(3, 3)$.

20. $A(7, 11), B(-9, -5)$.

In Problems 21 through 24 find an equation for the perpendicular bisector of the line segment joining the points A and B, using the method illustrated in Example 2. *If you recall from your earlier classes how to graph an equation, then in each problem make a sketch showing the line segment joining the points A and B and a few points on the perpendicular bisector.

21. $A(-3, 3), B(7, -7)$.

22. $A(-1, -3), B(3, 5)$.

23. $A(-4, 5), B(5, 6)$.

24. $A(1, 2), B(5, -7)$.

In Problems 25 through 28 find an equation for the circle with $C(a, b)$ as the center and R as the radius. Hint: The point $P(x, y)$ is on the circle if and only if $(x - a)^2 + (y - b)^2 = R^2$. Then simplify the equation that you obtain.

25. $C(3, -4), R = 5$.

26. $C(-4, 5), R = 9$.

27. $C(3, 2), R = 1$.

28. $C(4, -1), R = 11$.

A triangle is a right triangle if and only if the lengths of the sides satisfy an equation of the form $c^2 = a^2 + b^2$ for some choice of letters for the lengths. In the following problems determine if the given points are the vertices of a right triangle or not. In each case make an accurate drawing showing the triangle.

29. $A(1,3), B(4,2), C(-2,-6)$

30. $A(8,-3), B(-4,2), C(1,6)$,

31. $A(-3,1), B(1,-4), C(11,4)$

32. $A(-4,2), B(2,-1), C(0,-5)$.

33. $A(2,-6), B(-5,1), C(-1,5)$

A triangle is an isosceles triangle if two of the sides have the same length. In each of the following triangles determine whether the triangle is isosceles or not

34. $A(5,-2), B(6,5), C(2,2)$.

35. $A(2,1), B(9,3), C(4,-6)$

---

2. Graphs of Equations, An Overview

We begin with

Definition 2. The *graph* of an equation in two variables x and y is the collection of all points $P(x,y)$ for which the coordinates (x,y) make the equation true (satisfy the equation).

By computing and locating a number of these points, we form a sketch or picture of the graph. Although the picture or the sketch is not the same as the graph, it is usually sufficient for us to understand the graph that the sketch represents. Hence we are frequently quite satisfied with a sketch or a drawing and this chapter is devoted to making such sketches or drawings. Possibly, the reader has been sketching graphs of elementary equations for some time, and in this case he/she is aware that long and tiresome computations lie ahead. But blessed relief is in sight (using a graphing calculator or a computer), and before we give some specific examples of equations and their graphs we will discuss the use of these nice tools.

The invention of the graphing calculator and the computer is without doubt the greatest scientific achievement of the last quarter of the twentieth century. These tools make it childishly easy to obtain highly accurate sketches of graphs. This avoids the long and tiresome computations we mentioned in the previous paragraph.

Although a graphing calculator will be sufficient for most of the problems in these booklets, there are many problems where a computer may be necessary. Throughout this chapter and the rest of these booklets (and the rest of the student's scientific life) the student is expected to use whatever graphing calculator or computer may be available.

A remarkably fine program for a computer has been created at the United States Naval Academy by Professor Howard Penn and his many gifted associates. Because this program has been created at government expense it is available on the Naval Academy system and can be downloaded using following address

htt://www.usna.edu/MathDept/mpp

This program is named MPP (Mathematics Plotting Program).
A personal letter can be addressed to:

Professor Howard Penn
Mathematics Department
United States Naval Academy
572 Holloway Road
Annapolis, MD 21402

Since graphing calculators give us the graph of an equation without pain or discomfort, why should the student bother to make a few graphs "by hand" without the help of a graphing calculator?

Answer: The student must understand what is involved in graphing before he/she turns to this remarkable tool for help. The situation can be compared to using a car to travel. A person must as a baby learn to crawl before trying to walk. As one matures one must learn to walk, to run, and to ride a bicycle before one is ready to drive a car. This normal development demands a certain amount of effort and sweat, to complete the learning process. The same is true of graphing. A student must spend some time and effort making graphs by hand, before he/she can understand and appreciate what the graphing calculator can do.

In these booklets we assume that the student is at the very beginning of learning to graph. However, in all the exercises the problems are divided into two sets and clearly labeled (A) graphs to be sketched by hand, and (B) graphs to be sketched using a graphing calculator or a computer. If the reader has already suffered the required amount of labor and pain, then he/she can omit most or all of the problems in set (A), and use whatever tool is available for the problems in the set (B).

Exercise 2

(A)

1. Although the grahing calculator makes graphing childishly easy, it is highly desirable for the student to sketch a few graphs by hand. Name two other activities where machines work far better than the "hand" and yet it is highly desirable to learn to do the work "by hand", first before using the machine.

3. Graphs of Equations, Some Details

In this section we give a few examples, showing how to sketch the graph of an elementary equation by hand (meaning without the computer).

Example 1. Sketch the graph of the equation

$$x - y + 2 = 0\,. \tag{7}$$

Solution. We first solve the equation for y (if possible). If this is very difficult or impossible, then we should try to solve for x. If we cannot solve the equation for y or x, then we certainly need to use a computer program. If we solve Equation (7) for y we obtain

$$y = x + 2\,. \tag{8}$$

Since the graph is the collection of all pairs (x, y) for which the equation is true we select a few "nice" values of x and use Equation (8) to compute the associated y. Here the selection of "nice" values for x depends on the equation, but most problems are designed so that small integers should be selected for x. Here if $x = 1$, then Equation (8) gives $y = 3$, so the point $(1, 3)$ is a point of the graph. The work will go faster if we make a table and compute a few pairs. The results are shown in Table 1. The student should check each entry in the table. The selected x is in the first row, and the associated y is in the second row directly below x.

Table 1.

x	-3	-2	-1	0	1	2	3	4	—
y	-1	0	1	2	3	4	5	6	—

These points are shown in Figure 5. Since they all seem to lie on a straight line, we show this line in Figure 5 as the graph of $x - y + 2 = 0$ or $y = x + 2$. ▶▶

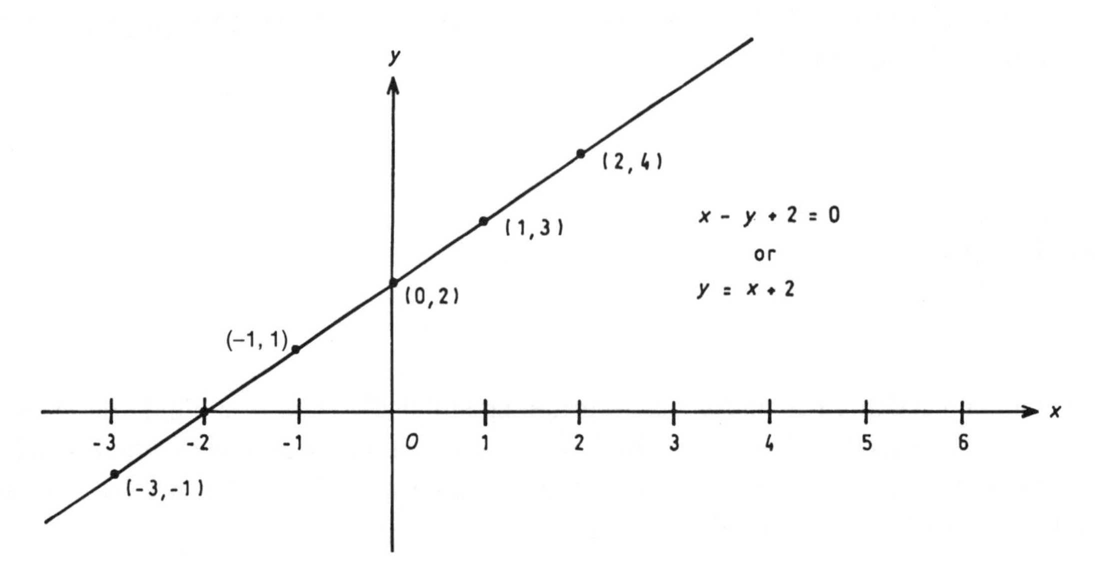

Figure 5.

Is the graph of $y = x + 2$ really a straight line as the sketch in Figure 5 seems to indicate? The answer is yes, but the situation is not a simple as one might at first suppose. We will give more details in the next section. In the meantime the student should observe that the graph of any equation of the form $Ax + By + C = 0$, seems to be a straight line and he is safe in assuming that indeed it is, if either A or B is not zero.

Example 2. Sketch the graph of $y = (x^2 - 2x - 4)/2$.

Solution. We have inserted the factor 1/2 to keep the points of the graph from getting too high. For example if $x = 6$, then (without the factor 1/2) $y = 20$, but with the factor 1/2 we find that $y = 10$. A table of values for $y = (x^2 - 2x - 4)/2$ is given in Table 2. The sincere reader should check each entry.

Table 2.

x	-3	-2	-1	0	1	2	3	4	5
y	$11/2$	2	$-1/2$	-2	$-5/2$	-2	$-1/2$	2	$11/2$

A sketch of the graph of $y = (x^2 - 2x - 4)/2$ is shown in Figure 6. The curve looks like a parabola and it is indeed. ▶▶

We will give many more details about the *Conic Sections* in the chapter titled "Parabolas, Ellipses and Hyperbolas". This will include a correct definition of a parabola, and the proof that the graph of $y = (x^2 - 2x - 4)/2$ really is a parabola.

This curve appears frequently in nature. For example the curve formed by water flowing from a drinking fountain is approximately a parabola (although it is upside down when

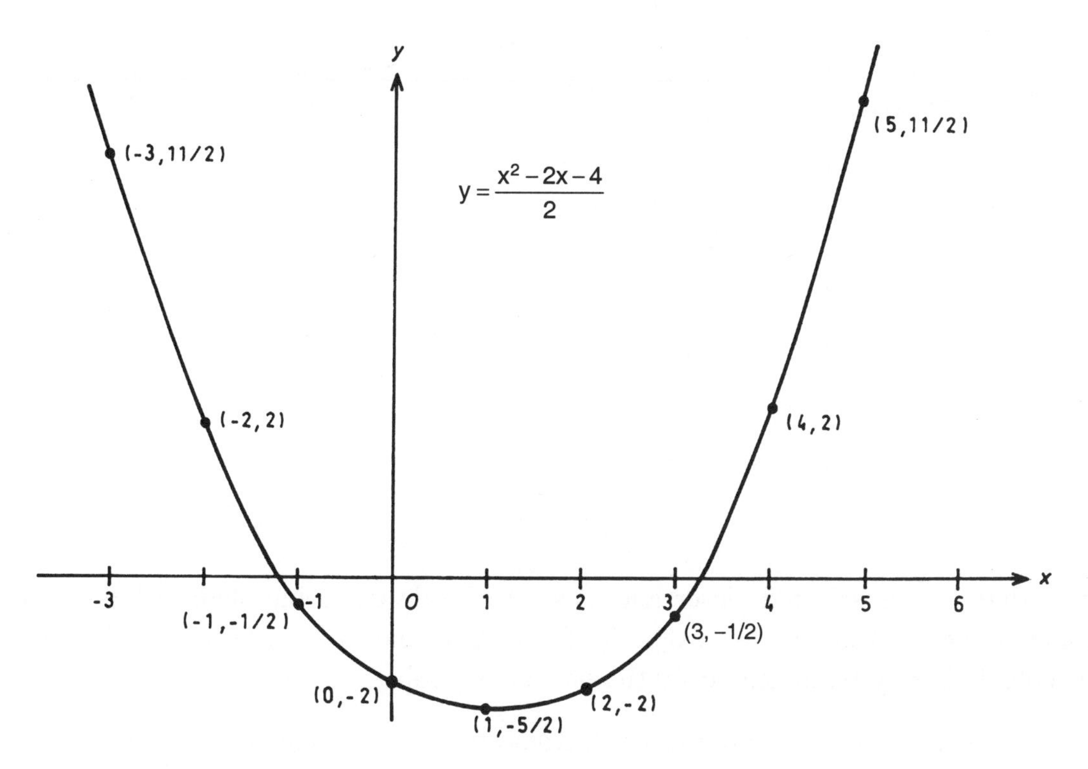

Figure 6.

compared with the graph in Figure 6). The same is true for the path of a stone when thrown upward and outward from a cliff.

Example 3. Sketch the graph of $y = 1/(x-2)$ for x between -2 and 6.

Solution. We first observe that when $x = 2$, y is undefined because the equation indicates a division by zero. Frequently, some intuitive mathematicians like to write $y = \infty$ in this situation. Whether we write ∞ or not, the division by 0 warns us to be careful and that we should select a few values for x that are near $x = 2$ the "bad" point, when we computing y. Such values are included in Table 3. As usual you should check all the computations indicated in that table. Naturally, the second row is computed using $y = 1/(x-2)$.

Table 3.

x	-1	1	1.5	1.75	2	2.25	2.5	3	5	8
y	$-1/3$	-1	-2	-4		4	2	1	1/3	1/6

A sketch of the graph is shown in Figure 7. ▶▶

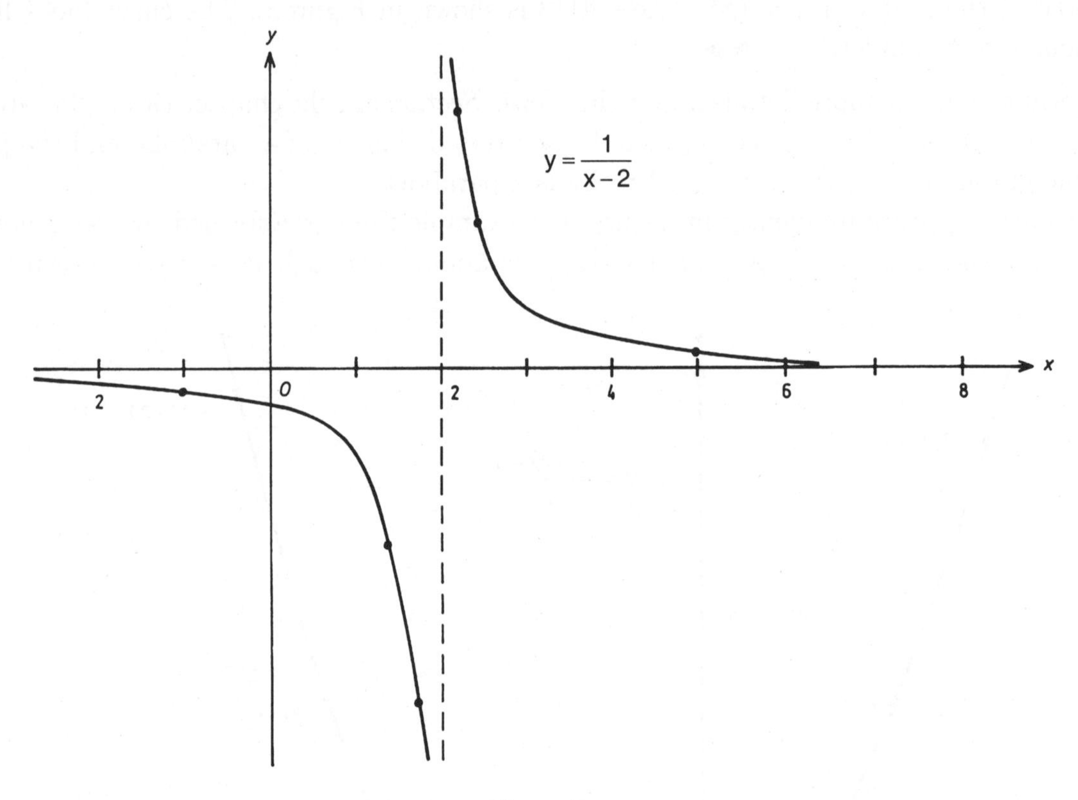

Figure 7.

Notice that as x is selected closer and closer to 2 (always being slightly larger than 2), y becomes larger and larger. For example if $x = 2.01$ then $y = 100$. If $x = 2.0001$, then $y = 10,000$. This may be indicated by the pair of symbols:

$$x \to 2^+ \quad \text{(read "}x\text{ approaches 2 from above", or "from the right")},$$

$$y \to \infty \quad \text{(read "}y\text{ approaches infinity")}.$$

The little plus sign located like an exponent over the 2 indicates that the values of x are to be slightly LARGER than 2. This pair usually appear in the form $y \to \infty$ as $x \to 2^+$ (read "y approaches infinity as x approaches 2 from above"). Frequently, these symbols are combined and the result is written as

$$\lim_{x \to 2^+} y = \infty ,$$

which is read "the limit of y as x approaches 2 from above (or from the right) is infinity".

In Example 3 (Figure 7) we also have the limit relations:

$$x \to 2^- \quad \text{(read "}x\text{ approaches 2 from below", or "from the left")},$$

$$y \to -\infty \quad \text{(read "}y\text{ approaches minus infinity")}.$$

For example, if $x = 1.999$ then $y = -1000$. In this example the second pair is combined to give $y \to -\infty$ as $x \to 2^-$ (read "y approaches minus infinity as x approaches 2 from below").

Just as before, these symbols are frequently combined and the result is written as

$$\lim_{x \to 2^-} y = -\infty ,$$

which is read "the limit of y as x approaches 2 from below (or from the left) is minus infinity".

These "limit" relations are not used very much in algebra, but are very important in graphing and in the study of calculus and many other advanced courses.

Exercise 3

As we mentioned in Chapter 7, Exercise 2, problems in a set labeled (A) should be done by hand (without using a computer). One should use a suitable computer program for problems in a set labeled (B). If a suitable program is not available it may be best to omit these problems. In all of these problems, the answer (graph) will be found on the monitor connected with your computer.

If the answer to a problem in a set labeled (A) is a graph, usually the graph will not be given here, because it is much easier for the student to use a suitable graphing calculator to obtain the answer. This includes the problems worked "by hand" where the student can check his/her work by using one of these new remarkable machines.

(A)

In Problems 1 through 8, the graph of the equation is a straight line. Find two points on the graph and draw the straight line joining those points. Then compute the coordinates of a few more points, and check that these new points seem to lie on the line that you drew. As an answer we will give the coordinates of two of the (infinitely many) points on the line.

1. $y = 2x + 5$. **2.** $y = -3x + 7$. **3.** $y = x - 4$. **4.** $y = -2x - 5$.

5. $3y = x + 6$. **6.** $4y = -x - 8$. **7.** $2x+5y+20 = 0$. **8.** $4x - 3y = 24$.

In Problems 9 through 17 sketch the graph of the given equation. These graphs are all parabolas, although we have not proved this fact. The proof will be given in Chapter 10.

9. $y = x^2$. **10.** $y = -x^2$. **11.** $y = x^2/4$.

12. $y = x^2 - 3$. **13.** $y = (x - 3)^2$. **14.** $y = (x + 3)^2$.

15. $y = (x - 3)^2 + 4$. **16.** $y = x^2 - 6x + 13$. **17.** $y = x(x - 6) + 9$.

In Problems 18 through 21 a circle is given by giving its center C and its radius R. In each case find an equation for the circle in the form $x^2 + y^2 + cx + dy + e = 0$.

18. $C(1, -2), R = 3$. **19.** $C(4, -3), R = 5$.

20. $C(1, -2), R = 10$. **21.** $C(2, -3), R = 6$.

In Problems 22 through 25 an equation is given. By completing the square find the graph.

22. $x^2 + y^2 - 4x + 2y - 20 = 0$ **23.** $x^2 + y^2 + 6x + 8y + 24 = 0$.

24. $x^2 + y^2 - 6x - 16y + 73 = 0$. **25.** $4x^2 + 4y^2 - 4x + 20y + 36 = 0$.

26. Find an equation for the set of all points P such that

$$|PA| = 3|PB|,$$

where A is the point $(3, 5)$ and B is the point $(3, -3)$.

The set of all of these points forms a circle. Find the center and radius of this circle.

27. Do Problem 26 if A is $(-1, 3)$ and B is $(7, 3)$.

(B)

To see the graphs for the problems in (B), use a suitable graphing program and view the graphs on your monitor

In Problems 28 through 33, sketch the graph of the equation.

28. $y = |x|$. **29.** $y = |x - 5|$. **30.** $y = |x + 6|$.

31. $y = |x| - 5$ **32.** $|x| = y^2$. **33.** $x = |y|$.

In Problems 34 through 38, sketch the graphs of the given equation for each of the given values of c. In each problem put all of the graphs on the same coordinate system.

34. $x + y = c$, for $c = 1, -1, 5, -5, 10$, and -10.

35. $2y = x + c$, for $c = 1, -1, 5, -5, 10$, and -10.

36. $y = x^2/c$, for $c = 1, 2, 4, 8, -1, -2, -4$, and -8.

37. $x^2 + y^2 = c$, for $c = 4, 6, 8$, and 10.

38. $y = x^c/10$, for $c = 2, 3, 4, 5$.

In Problems 39 through 49 sketch the graph of the given equation.

39. $y = x(x-4)^2/5$.

40. $y = x(x^2-4)/5$.

41. $y = x(x-2)(x-4)(x-6)$.

42. $y^2 = x^3/10$.

43. $x^4 + y^4 = 81$.

44. $y = |x-1| + |x+1|$.

45. $y = |x^2 - 4|$.

46. $y = |x(x-3)(x+3)|$.

47. $y = x - 1 - |x-2| + |x-5|$.

48. $y = 3 + x - |x-2| + |x-5| - |x+2|$.

49. $y = (|x^2-1| - |x^2-25|)/5$.

In Problems 50 through 55, sketch the graph of the polynomial for x in the interval $[-1, 1]$. Observe that these polynomials are of degree 2, 3, 4, 5, 6, and 7 respectively. Also notice that all the coefficients are integers and that the maximum value of $|y|$ in that interval is 1. Tchebychev *PROVED* that of all polynomials with integer coefficients, the ones with the smallest maximum $|y|$ in $[-1, 1]$ are just the ones given in Problems 50 through 55.

50. $y = 2x^2 - 1$.

51. $y = 4x^3 - 3x$.

52. $y = 8x^4 - 8x^2 + 1$.

53. $y = 16x^5 - 20x^3 + 5x$.

54. $y = 32x^6 - 48x^4 + 18x^2 - 1$.

55. $y = 64x^7 - 112x^5 + 56x^3 - 7x$.

We obtain a very pretty picture if we put all of the graphs in Problems 50 through 55 on the same coordinate system.

---

4. The Straight Line

One expects this section to begin with a definition of a straight line. Actually this is far more difficult than you may suppose. From early childhood we have seen many examples of straight lines. The rulers, the edges of buildings, the edge of a book, or the edge of a sheet of paper, are all examples of straight lines that we accept without thinking. But if we are doing mathematics this common experience is not sufficient as a definition. The great Euclid starts with the definition (naturally in Greek) that a straight line is "a line which lies evenly with the points on itself". This is taken from the books by Sir Thomas L Heath, 1908, who is considered the final authority on the works of Euclid. Modern critics jump on poor Euclid for writing such nonsense, but as we will see in a moment, Euclid was not as stupid as these critics want us to believe. But first, a word about definitions in general. To define a word, we must use other words, and it is assumed that the reader knows the meanings of the other words used. If not, then these words must be defined, using still other words. Following this process backwards to the beginning, we are forced to admit that some items (elements, words) must be taken as basic and undefined. The true character of the undefined element is given by its properties (the axioms that the element must satisfy). The modern view is that "point, line, and plane" are undefined elements. Now Euclid's critics are wrong because if we consider the great works that Euclid collected (or created himself) it is obvious that he must have known that some elements must be undefined. So why didn't Euclid say so? Recall that Euclid was writing several thousands of years ago, and that he probably wanted his work to be studied by people at all levels of development, and his "faulty" definition of a line was inserted merely to satisfy the beginning student (and was never aimed at today's stodgy scholars). Other (so-called) errors of Euclid can be explained in the same way. Euclid was much smarter than his many pretentious critics.

In this book we will actually give a definition of a straight line using the rectangular coordinate system. In this way the reader is left with a feeling that we have overcome the difficulty. Actually we have buried the difficulty so that it is hard to find, but in so doing we believe that the reader will be more satisfied with the presentation given here.

Definition 3 (Slope). The *slope* of the line segment joining the two points $P_1(x_1, y_1)$ and $P_2(x_2, y_2)$ is denoted by m and is given by the formula

$$m = \frac{y_2 - y_1}{x_2 - x_1}, \tag{8}$$

if $x_1 \neq x_2$ (no division by zero). If $x_1 = x_2$ the slope is undefined (although the symbol $m = \infty$ is often used).

Notice that the word "line" crept into the definition. This is not an error. The definition of slope depends only on the two points P_1 and P_2 and the "line" is really not involved. (Think about it.)

Example 1. Find the slope for the two points shown in Figure 8.

Solution. We may select either point to be P_1. In this case we choose $(2, 1)$ for P_1 and hence $(5, 3)$ must be P_2. Then Equation (8) gives

$$m = \frac{3-1}{5-2} = \frac{2}{3}. \tag{9}$$ ▶▶

Observe that the answer would also be $2/3$ if we selected $(5, 3)$ for the first point. The numerator in (9) is called the *rise* and the denominator is called the *run*. The reason for these names becomes clear when we look at Figure 8. Of course, the rise may be a negative number. For example, a rise of -8 units would be called a *fall* of 8 units.

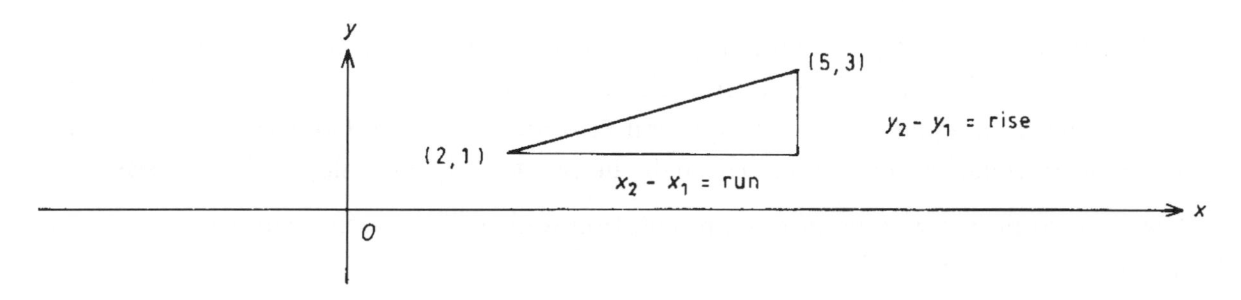

Figure 8.

Now we are in a position to define a straight line. Briefly a collection of points form a straight line if every pair of points in the set gives the same number in Equation (8) for the slope.

Thus a line is a collection of certain points.

Definition 4 (A Line with a Given Slope). Suppose that m is a given number and $P_1(x_1, y_1)$ is a given point. Then the straight line through P_1 with slope m is the point P_1 plus all the points $P(x, y)$ for which

$$\frac{y - y_1}{x - x_1} = m\,, \quad x \neq x_1\,. \tag{10}$$

Thus the line consists of the point P_1 together with all points P for which the pair P_1 and P has slope m.

Notice that the language seems somewhat strained because P_1 is treated as a special point. But this strain is necessary because if we try to replace (x, y) by (x_1, y_1) in Equation (10) we face a forbidden division by zero, namely the denominator becomes $x_1 - x_1 = 0$. But we want P_1 to be a point on the line.

Notice that a straight line is defined by an equation. However in the future we will no longer refer to the definition, but instead say that Equation (10) is an equation of a straight line.

Example 2. Find an equation for the line through $P_1(3, 4)$ with slope $1/2$, and simplify the result.

Solution. We substitute the appropriate numbers in equation (10) and find that

$$\frac{y - 4}{x - 3} = \frac{1}{2} \tag{11}$$

is an equation for the line. But we are allowed to convert (11) into other equivalent equations. Elementary multiplications will give

$$2(y - 4) = 1(x - 3)\,. \tag{12}$$

Continuing with other elementary manipulations we will find

$$2y - 8 = x - 3 \tag{13}$$

$$2y - x - 5 = 0 \tag{14}$$

$$y = \frac{1}{2}x + \frac{5}{2}\,. \tag{15}$$

Since Equations (12), (13), (14) and (15) are all equivalent we can say that any one of them is an equation for the line that goes through $(3,4)$ and has slope $1/2$. Notice that all of the steps in going from equation (11) to equation (15) are reversible. Consequently, we can say that (15) is an equation for the line through the point $(3,4)$ with slope $1/2$. ▶▶

It is time to fill a few gaps in our presentation and to introduce a suitable terminology. How about vertical lines? This is

Definition 5 (Vertical line). The set of all points $P(x,y)$ for which

$$x = c, \tag{16}$$

where c is a constant, is a *line*, a *vertical line* through the point $(c,0)$.

These lines were omitted in Definition 4. Further, in Definition 4, the point $P_1(x_1,y_1)$ was treated separately to avoid a division by zero. But multiplication will transform Equation (10) into

$$\boxed{y - y_1 = m(x - x_1)\,.} \tag{17}$$

Here, the division by zero that occurs when $x = x_1$ has been replaced by multiplication by zero, which is well-defined. For obvious reasons Equation (17) is called the *point-slope form* for the equation of a straight line.

Suppose that we are given two points on the line rather than one point and the slope. Then we use Equation (9) to compute the slope and substitute this in Equation (17). This gives

$$\boxed{y - y_1 = \frac{y_2 - y_1}{x_2 - x_1}(x - x_1)\,.} \qquad x_1 \neq x_2\,. \tag{18}$$

Equation (18) is called the *two-point form* for the equation of a straight line.

Example 3. Find an equation for the line through the two points $(2,1)$ and $(5,3)$ given in Example 1 (see Figure 8) and simplify.

Solution. Using these numbers in Equation (18) we have

$$y - y_1 = \frac{y_2 - y_1}{x_2 - x_1}(x - x_1)\,, \quad \text{or} \quad y - 1 = \frac{3-1}{5-2}(x - 2)\,.$$

To simplify we multiply both sides by 3 obtaining

$$3(y-1) = 2(x-2)\,, \quad \text{or} \quad 3y - 3 = 2x - 4\,, \quad \text{or} \quad 3y = 2x - 1\,.$$

This last equation may be considered as the simplest form. But, we may divide by 3 to obtain another simple form, where y stands alone,

$$y = \frac{2}{3}x - \frac{1}{3}\,. \tag{19}$$ ▶▶

It is often a good idea to check your work by substituting the coordinates of the given points into the final result to see if they really satisfy the final equation. Here we leave it for the student to check that $(2,1)$ and $(5,3)$ satisfy Equation (19).

We recall from Example 1, that the slope of the line joining the given points is $2/3$, and hence this is also the slope of the line in Example 3. Please notice that this same number $2/3$ is also the coefficient of x in (19). This is not an accident. Indeed, we will soon prove that any equation of the form $y = Mx + B$ is the equation of a straight line for which the slope $m = M$. First we need the definition of the intercepts of a line on the axes.

Definition 6 (The Intercepts). If a line meets the x axis at the point $(a, 0)$, then a is called the *x-intercept* of the line. If it meets the y-axis at the point $(0, b)$, then b is called the *y-intercept* of the line.

Example 4. Find the two intercepts of the line $4y - 3x = 12$.

Solution. Observe that we have not proved that the graph of this equation is a line. Nevertheless, we can find where it meets the two axes. If we put $x = 0$ in $4y - 3x = 12$, we find that $4y = 12$, or $y = 3$. Hence, $(0,3)$ is a point of the graph and therefore $b = 3$ is the y-intercept. If we put $y = 0$ we obtain $-3x = 12$, or $x = -4$. Hence, the point $(-4,0)$ is a point of the graph and therefore $a = -4$ is the x-intercept. The graph of $4y - 3x = 12$ together with the two intercepts is shown in Figure 9. ►►

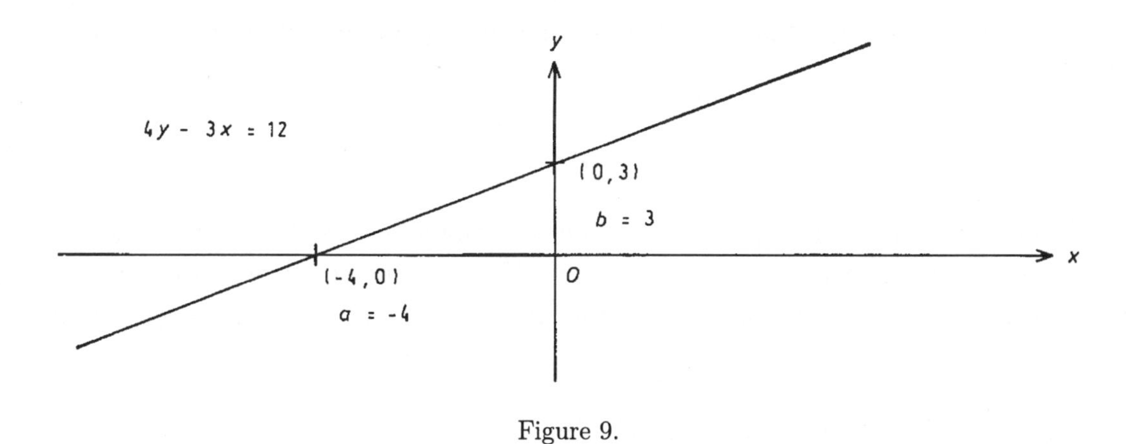

Figure 9.

Theorem 4. *For any two numbers m and b, the graph of*

$$y = mx + b \tag{20}$$

is a line. The number m is the slope and b is the y-intercept.

Equation (20) is called the *slope-intercept* form of the equation of a line.

Proof of Theorem 4. According to our definition any two points on the graph of (20) must give the same slope. Suppose that $P_1(x_1, y_1)$ and $P_2(x_2, y_2)$ are any two points on the graph of (20). Then the coordinates of these points satisfy Equation (20) and hence the equations

$$y_2 = mx_2 + b\,, \tag{21}$$

and

$$y_1 = mx_1 + b\,, \tag{22}$$

are both true at the same time (simultaneously). If we subtract Equation (22) from Equation (21), the term b disppears. When we factor m, we obtain $y_2 - y_1 = m(x_2 - x_1)$. If the two points P_1 and P_2 are different, the last factor is not 0 and division gives us

$$\frac{y_2 - y_1}{x_2 - x_1} = m\,.$$

Thus every pair of points on the graph of (20) gives the same slope m. Hence the graph of (20) is a line with slope m. If we put $x = 0$, we obtain $y = b$, so $(0, b)$ is on the line and b is the y-intercept. ■

Theorem 5. *The graph of*

$$Ax + By = C \tag{23}$$

is either a line, or is every point in the plane, or has no points.

Proof. Suppose $B = 0$ and $A \neq 0$. Then dividing both sides of (23) by A gives the equivalent equation $x = C/A$, the equation of a vertical line. If $B \neq 0$, then dividing both sides by B will lead to

$$y = -\frac{A}{B}x + \frac{C}{B}\,, \tag{24}$$

and by Theorem 4, this is the equation of a line with slope $-A/B$ and y-intercept C/B. Now if both A and B are 0, then (23) becomes $0 = C$. If $C = 0$ the equation $0 + 0 = 0$ is satisfied by the coordinates of every point of the plane. If $C \neq 0$, then, $0 = C$ is never true and the graph has no points. ■

Example 5. What is the graph of $7x + 19y = 37$?

Solution. This equation is equivalent to $y = -7x/19 + 37/19$. Hence by Theorem 5 the graph is a line with slope $-7/19$ and y-intercept $37/19$. ▶▶

Example 6. What is the graph of $xy + 2 = x + 2y$?

Solution. This equation is equivalent to $xy - x - 2y + 2 = 0$. Students with sharp eyes will notice that this expression can be factored, so the given equation is equivalent to

$$(x - 2)(y - 1) = 0\,.$$

This product is zero if and only if at least one of the factors is zero. So either $x - 2 = 0$, or $y - 1 = 0$ (or both). Thus the graph is composed of two straight lines, a vertical one at $x = 2$ and a horizontal one at $y = 1$. ▶▶

Exercise 4

(A)

In Problems 1 through 8 find the slope of the line joining the two points given. Locate the points on a coordinate system and draw the line segment joining the two points.

1. $(2,3)$ and $(4,-7)$.

2. $(-4,-3)$, and $(-6,7)$.

3. $(5,7)$, and $(-1,-6)$.

4. $(1,-4)$, and $(-2,-5)$.

5. $(-2,12)$, and $(3,17)$.

6. $(-4,-5\sqrt{3})$, and $(6,5\sqrt{3})$.

7. $(-3,2)$, and $(-3,-4)$.

***8.** $(-7/2,3/4)$, and $(10/3,8/3)$.

In Problems 9 through 16 find the equation for the line through the two given points. Put your answer in the form $y = mx + b$ (if this is possible).

9. The points of Problem 1.

10. The points of Problem 2.

11. The points of Problem 3.

12. The points of Problem 4.

13. The points of Problem 5.

14. The points of Problem 6.

15. The points of Problem 7.

***16.** The points of Problem 8.

17. In Problems 9 through 16, check your answer $y = mx + b$ by showing that the coordinates of both of the given points satisfy the equation that you found.

In Problems 18 through 23 find the slope and y-intercept for the given line.

18. $3x + 5y = 35$.

19. $5x + 3y + 36 = 0$.

20. $x + y + 7 = 9$.

21. $7x = 3y - 57$.

22. $\pi x + 2\pi y = 3$.

23. $0 = 7x + 9y + 60$.

24. If $ab \neq 0$ prove that the graph of $x/a + y/b = 1$ has the intercepts a and b on the x- and y-axes respectively.

In Problems 25 through 28 give an equation of the form $Ax + By = C$ for the line with the given x- and y-intercept respectively.

25. $5,-4$.

26. $1,2$.

27. $-2,7$.

28. $-1/2,1/6$.

In Problems 29 through 35 find the graph of the given equation.

***29.** $y^2 = 4x^2$.

***30.** $(y-3)^2 = (x+1)^2$.

***31.** $y^2 - 4y + 4 = x^2 - 2x + 1$.

***32.** $y = 1 + x + |x|$. Hint: Consider two cases, $x \geq 0$, and $x < 0$.

***33.** $y = 2 + |x|$.

***34.** $y = 2 - |x|$.

***35.** $y = 3 + |x - 2|$.

36. For students who have had some trigonometry. Let α (Greek letter "alpha") be the angle that a line makes with the positive x-axis. If the line is not vertical show that the slope m is the same as $\tan\alpha$.

In Problems 37 through 40 locate the points P, Q, and R, on a rectangular coordinate system. Do they seem to be collinear (fall on a straight line)? Then determine if they are collinear by computing the slopes of PQ and PR. Check your work by finding the slope of QR.

37. $P(-1,-2)$, $Q(5,-5)$, $R(-5,0)$.

38. $P(2,4)$, $Q(-3,2)$, $R(-13,-2)$.

39. $P(-6,8)$, $Q(4,2)$, $R(12,-3)$.

40. $P(-2,5)$, $Q(-12,-11)$, $R(6,18)$.

---

5. More About Straight Lines

Given the equations of two lines, can we decide if they are parallel lines or perpendicular lines by looking at their equations?

Definition 7 (Parallel Lines). Two lines are *parallel*, if they do not meet (there is no point that is simultaneously on both lines).

When we say two lines we mean two distinct lines. This eliminates the case where both lines are the same. But it is convenient to regard a line as parallel to itself. The beginning student is wise to ignore this very special case in which the two lines are the same. The professional mathematician likes to worry about it. However, if the lines coincide, then they are said to be parallel.

First we look at the trivial case of two vertical lines. If the two lines have the equations $x = a_1$ and $x = a_2$ respectively, then the two lines are the same if $a_1 = a_2$, and hence parallel. If $a_1 \neq a_2$, then the lines $x = a_1$ and $x = a_2$ are distinct and parallel. If the lines are not vertical, then we have

Theorem 6. *If two lines have equations*

$$y = m_1 x + b_1 , \tag{25}$$

and

$$y = m_2 x + b_2 , \tag{26}$$

respectively, then the lines are parallel if and only if $m_1 = m_2$.

Proof. If $m_1 = m_2$ and $b_1 = b_2$ then the two lines coincide and in this trivial case the lines are parallel. Now assume that the two lines are distinct. Then in Equations (25) and (26), either $m_1 \neq m_2$ or $b_1 \neq b_2$ or both. If we assume that there is a point where the two lines meet (intersect), then the coordinates (x, y) of that point satisfy both Equations (25) and (26). If we subtract the first equation from the second one we arrive at

$$0 = m_2 x + b_2 - m_1 x - b_1 \quad \text{or} \quad 0 = (m_2 - m_1)x + b_2 - b_1 . \tag{27}$$

Now, if $m_2 \neq m_1$, then the second equation in (27) can be solved for x, and this will give an associated y so there *IS* a point of intersection of the two lines. Thus, the lines are not parallel. So for parallel lines it is necessary that $m_2 = m_1$. In this case the second equation in (27) gives $0 = b_2 - b_1$. If $b_2 = b_1$, then the two lines coincide. If $b_2 - b_1 \neq 0$, then the assumption that the two lines intersect leads to a contradiction that $0 = b_2 - b_1 \neq 0$. Hence if $m_1 = m_2$ the two lines do not meet and are therefore parallel. ■

For example, the two lines with equations $y = (3-\pi)^2 x + 57$ and $y = (3-\pi)^2 x - \sqrt{23/19}$ are parallel. It is not ncessary to locate any points on the graphs of the two equations.

Definition 8 (Perpendicular Lines). Two lines are *perpendicular* if they meet at right angles.

We first rule out the trivial case in which one line is vertical and one line is horizontal. Thus the two lines $x = a$ and $y = b$ are obviously perpendicular and need no further discussion.

Theorem 7. *If two lines have the Equations* (25) *and* (26), *then they are perpendicular if and only if*

$$\boxed{m_1 m_2 = -1 \quad \text{or} \quad m_2 = -\frac{1}{m_1} .} \tag{28}$$

Proof. First notice that b_1 (or b_2) is not involved in Equation (28). Then by changing b_1 (or b_2) we can move the corresponding line up or down without changing the slope of either line, and hence not changing the angle of intersection of the two lines. So we can assume, without loss of generality, that the lines meet above the x-axis as shown in Figure 10. This movement upward of the lines does no damage to the theorem or to the proof. Let A, B, and C be the *positive* distances shown in the figure. Then the angle at Q is a right angle if and only if the triangle PQR is a right triangle and by the Pythagorean Theorem, if and only if

$$|PQ|^2 + |QR|^2 = |PR|^2 ,$$

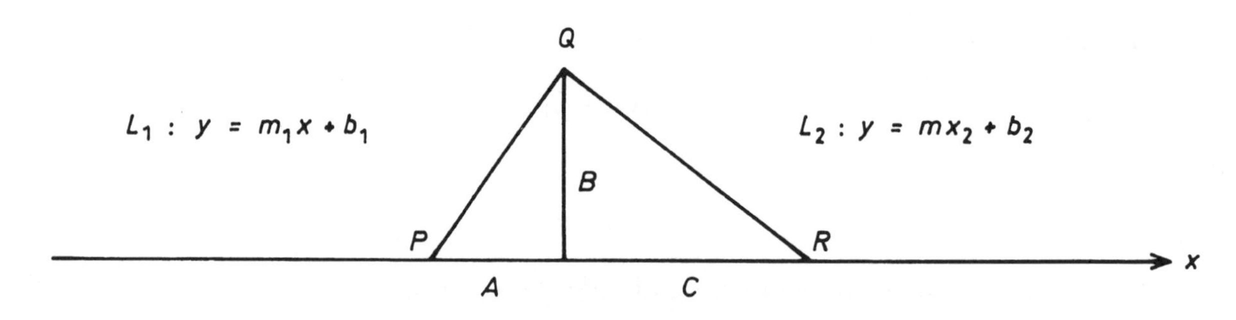

Figure 10.

or

$$(A^2 + B^2) + (B^2 + C^2) = (A + C)^2 .$$

On expansion and cancellation this gives

$$2B^2 = 2AC ,$$

or

(29) $$\frac{B}{A}\frac{B}{C} = 1 .$$

But $B/A = m_1$ and $B/C = -m_2$ (watch the minus sign) so Equation (29) gives Equation (28).

■

Example 1. Are the two lines $y = 3x + \pi$ and $3y + x - \sqrt{37} = 0$ perpendicular?

Solution. Yes. $m_1 = 3$ and $m_2 = -1/3$ and these numbers satisfy equation (28). ▶▶

Example 2. Find an equation for the line perpendicular to the line $y = 5x + 713$, and passing through the point $(2, -4)$.

Solution. Since $m_1 = 5$, Equation (28) tells us that $m_2 = -1/5$. Thus $y = -x/5 + b$ or $5y = -x + 5b = -x + c$ (using c in place of $5b$). We substitute the coordinates $(2, -4)$ in this last equation and obtain $5(-4) = -2+c$ in place of $5y = -x+c$. Therefore $c = -20+2 = -18$. The answer is $5y = -x - 18$, or (if you prefer) $x + 5y + 18 = 0$. ▶▶

Example 3. Find the point of intersection of the two lines

(30) $$2x - 3y = -5 ,$$

and

(31) $$x + 6y = 20 .$$

Solution. Let (x, y) be the coordinates of the point of intersection. For these values of x and y both Equations (31) and (32) are satisfied (true). Thus we have a pair of *simultaneous equations*. A thorough discussion of this topic will be given in a later chapter, but for the present we can learn how to handle this simple example. We multiply Equation (31) by 2 and obtain

$$2x + 12y = 40\,. \tag{32}$$

Then, if we subtract Equation (32) from Equation (30), the $2x$ term will disappear (that is why we multiplied (31) by 2). This gives

$$-15y = -45\,. \tag{33}$$

(Recall that subtraction means the addition of the negative, so multiply Equation (32) by -1 and add the result to Equation (30).)

Now, Equation (33) gives $y = -45/(-15) = 3$. If we use $y = 3$ in Equation (31) we find that $x = 20 - 6y = 20 - 18 = 2$. Thus the two lines intersect at the point $(2, 3)$. ►►

We can check our solution by using these numbers in Equations (30) and (31). From (30) we have

$$2x - 3y = 2(2) - 3(3) = 4 - 9 = -5\,,$$

and from (31) we have

$$x + 6y = 2 + 6(3) = 2 + 18 = 20\,.$$

Thus $(2, 3)$ must be the (only) point of intersection.

Suppose that we want to find the midpoint of a line segment. The formulas for the coordinates are given in

Theorem 8. *If (x_M, y_M) are the coordinates of the point M that is the midpoint of the line segment from $P_1(x_1, y_1)$ to the point $P_2(x_2, y_2)$, then*

$$\boxed{x_M = \frac{x_1 + x_2}{2}\,, \quad \text{and} \quad y_M = \frac{y_1 + y_2}{2}\,.} \tag{34}$$

Isn't this a simple set of equations? For a coordinate of the midpoint one merely computes an average of the corresponding coordinates of the end points of the line segment.

Proof. We use similar triangles. From Figure 11 it is clear that if M is the midpoint of the segment P_1P_2, then the two triangles

$$P_1MR \quad \text{and} \quad P_1P_2S$$

are similar with the proportionality factor $1/2$. Hence we have the equations

$$x_M - x_1 = \frac{1}{2}(x_2 - x_1) \quad \text{and} \quad y_M - y_1 = \frac{1}{2}(y_2 - y_1)\,. \tag{35}$$

We leave it for the reader to show that the equations in the set (35) give the equations in the set (34). ■

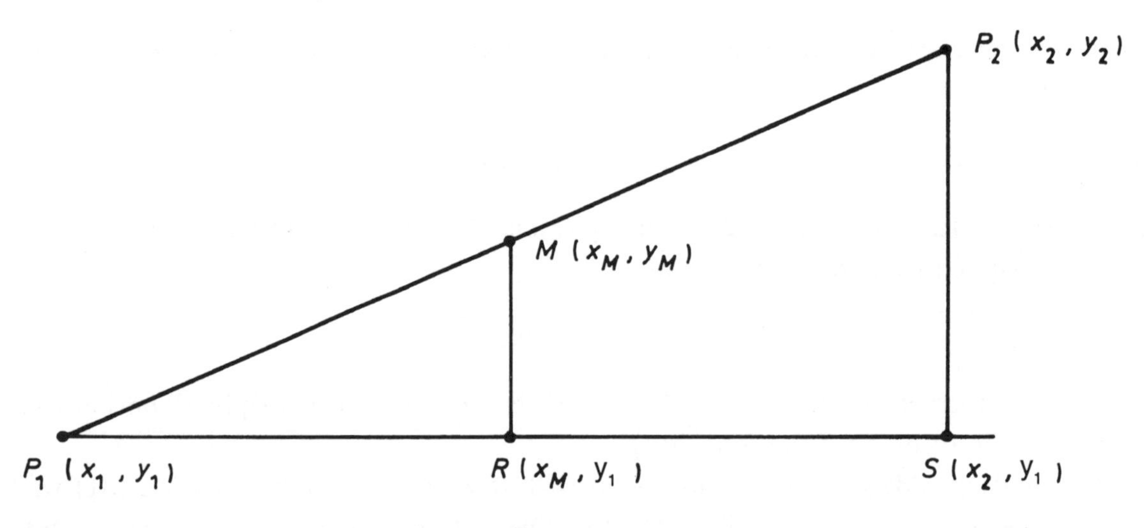

Figure 11.

In Exercise 5 an alternate proof of Theorem 8 will be outlined. This proof will use the distance formula and the slope of the line segment P_1P_2.

Example 4. Find the midpoint of the line segment P_1P_2 where P_1 is $(-3, 7)$ and P_2 is $(13, -10)$.

Solution. From Equation set (34) we have $x_M = (-3 + 13)/2 = 5$ and $y_M = (7 - 10)/2 = -3/2$, so M is $(5, -3/2)$. ►►

Exercise 5

(A)

In Problems 1 through 6 a point P and a line L are given. Find an equation for the line through P and parallel to L.

1. $P(5, -5)$, $L : y = x + 10$.

2. $P(0, 11)$, $L : 2y = x - 7$.

3. $P(0, 0)$, $L : 3y + x = \pi^2$.

4. $P(-1, -1)$, $L : 5y - 2x = 9$.

5. $P(100, 200)$, $L : x - 37 = 0$.

6. $P(3/4, 5/6)$, $L : 7x - 6y = 0$.

In Problems 7 through 12 find an equation for the line through P and perpendicular to L, for the points and lines in the first 6 problems.

7. For Problem 1.

8. For Problem 2.

9. For Problem 3.

10. For Problem 4.

11. For Problem 5.

12. For Problem 6.

In Problems 13 through 17 find the point of intersection (if there is one) for the pair of given lines.

13. $2y = x - 5$, and $3y = -2(x+2)$.

14. $3y + 5x = 7$, and $y = 4x - 9$.

15. $3x - 2y = 12$, and $2y = 4 - 5x$.

16. $3y = 2x + 1$, and $3y + 4x = 25$.

17. $\dfrac{y}{3} + \dfrac{x}{2} = \dfrac{7}{6}$ and $\dfrac{3y}{2} = \dfrac{2x}{3} + \dfrac{7}{3}$.

***18.** A theorem from geometry states that an angle inscribed in a semicircle is a right angle. Prove this theorem using Theorem 7. Hint: Let the circle be $x^2 + y^2 = r^2$ and let $A(-r, 0)$ and $B(r, 0)$ be the end points of a diameter. If $P(x, y)$ is any point on the upper half of the circle, prove that APB is a right angle.

In Problems 19 through 24 find the midpoint of the line segment P_1P_2.

19. $P_1(4, 10)$, $P_2(12, 20)$.

20. $P_1(5, 11)$, $P_2(7, 15)$.

21. $P_1(-7, 3)$, $P_2(3, -11)$.

22. $P_1(-1, -2)$, $P_2(-9, 10)$.

23. $P_1(1/2, -5)$, $P_2(1/3, 6)$

24. $P_1(3, -8)$, $P_2(-10, 5)$.

***25.** Prove Theorem 8, by showing that if M has the coordinates given in equation set (34), then $|P_1P_M| = |P_MP_2|$, and that the slope of the line segment P_1P_M is the same as that of the line segment P_1P_2. Explain why these two results prove Theorem 8.

Chapter 9

Inequalities

1. Some Notations

We have met some inequalities previously in an informal way. In this chapter we begin a more detailed study of the subject.

What is a positive number? From our earlier work we recall that we started with the positive integers (the counting numbers). Then we defined the positive rational numbers as ratios of positive integers. Finally, the positive real numbers were explained as limits of sequences of positive rational numbers. Here we must exclude 0, because it is not a positive number, although it is the limit of the sequence $[1/n] = 1, 1/2, 1/3, \ldots$. Thus we know the correct definition of a positive number, and further we should feel that we really understand the concept. But for practical work, it still may be very difficult to recognize a positive number. For example if $x = \sqrt{10} - \pi - 2/97$, it is hard to know if x is a positive number or not, without the help of a calculator (it really is positive). Before going further we must say a word about notation.

The symbols $<$, $\leq$, $>$, $\geq$ have been used previously. In this chapter we will use the symbol P to denote the set of all positive numbers (rational or irrational). The symbol $\in$ represents the phrase "is a member of", or "belongs to" so that the symbols $x \in P$ means x is an element of the set P. This is a fancy way of saying that x is a positive number, and is indeed much shorter. Most of the time we will not use this fancy symbol $\in$, but instead we will write "x is in P", because it is almost as short and we don't want to burden the student with too many new symbols. But you will surely meet the symbol $\in$ if you read other books on mathematics. The symbols $a < x < b$ mean that x is a number between a and b but is not equal to either a or b. The symbols $a \leq x \leq b$ mean that x is a number between a and b, and also may be either a or b. The set of all x such that $a < x < b$ is denoted by (a, b) and is called an *open interval.* The set (a, b) is conceptually different from the condition $a < x < b$, since the first is a set and the second is a restriction. Similarly the set of all x such that $a \leq x \leq b$ is denoted by $[a, b]$ and is called a *closed interval.* The open interval (a, b) looks like the point with coordinates (a, b) but one can tell what (a, b) means by the context (the way it is used). We can also have a *half open* (or *half closed* interval) $[a, b)$ where a is in the interval and b is not. Similarly a is not in the half open interval $(a, b]$ but b is in that interval. The numbers a and b are called *end points of the interval.*

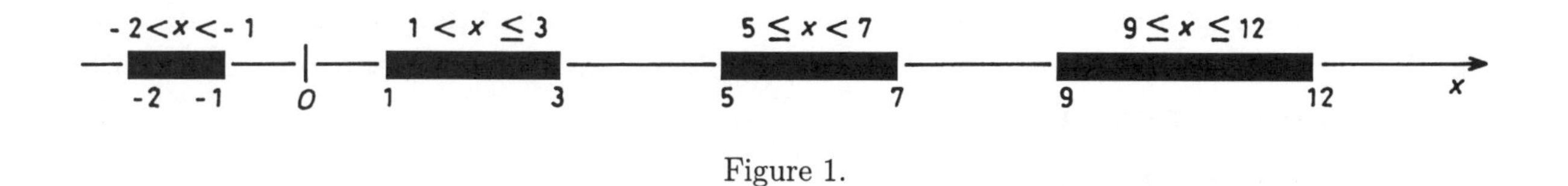

Figure 1.

Sometimes it is convenient to make a picture showing these intervals. The custom is to "shade" the points that belong to the interval and to use the same symbols [,], (,) to indicate if the end points are in the interval or not. A few intervals are shown in Figure 1.

The set $X \cup Y$ is the *union* of the sets X and Y, and consists of all the elements that are either in the set X or in the set Y or in both.

Example 1. $[-4, 5) \cup [5, 7]$ is the interval (set) $[-4, 7]$. But the set $[-4, 3) \cup [5, 7]$ cannot be rewritten as a single interval.

The set $X \cap Y$ is the *intersection* of the sets X and Y and consists of all the elements that are in both the set X and the set Y.

Example 2. $[2, 5] \cap [4, 10]$ is the interval (set) $[4, 5]$.

Example 3. $[2, 5] \cap [7, 10]$ is called the empty set because it does not have any points (elements). Frequently one uses the symbol ϕ for the empty set because it resembles a zero with a line through it. There are other fancy symbols for the empty set but ϕ is good enough for us. As an example $(2, 5) \cap (5, 7) = \phi$.

From our examples and problems one might think (erroneously) that $\cap$ and $\cup$ apply only to intervals. The theory of sets is a vast field that applies to all types of sets, but we want to treat only those items that may be useful in algebra.

There are a few other symbols that are very popular, and since the reader may meet them in other books we should list them here. We will meet some of these symbols again in Chapter 12.

The symbol $\subset$ is read "is contained in". Naturally $A \subset B$ means that the set A is contained in the set B. Thus every element of A is also an element of B. The compound symbol $A \subseteq B$ means that either A is contained in B or A is equal to B (the two sets are the same).

For example $[3, 4] \subset [1, 7)$.

The symbol $\supset$ is read "contains". Naturally, $A \supset B$ means that the set A contains the set B. Thus every element of B is also an element of A. The compound symbol $A \supseteq B$ means that either the set A contains the set B or A is equal to B (the two sets are the same).

The symbol $\Rightarrow$ is read "implies". Thus $A \Rightarrow B$ means that by some logically correct argument it can be proved that if the statement A is true then the statement B is also true. For example

$$n \text{ is a prime number greater than } 2 \Rightarrow n \text{ is a positive odd integer}\,.$$

The symbol $\Leftarrow$ is read "is implied by". It is similar to $\Rightarrow$ but the logical argument goes in the opposite direction. For example

$$\text{I am alive} \Leftarrow \text{I am in good health}\,.$$

The compound symbol $\Leftrightarrow$ may be read "if and only if" and the symbols $A \Leftrightarrow B$ means that A implies B and B implies A. Thus either statement is true if and only if the other statement is true. For example

$$x^2 - 3x + 2 = 0 \quad \Leftrightarrow \quad x = 1 \text{ or } x = 2 \quad \Leftrightarrow \quad x \in \{1, 2\}.$$

We will have much more to say about sets in Chapter 12.

Exercise 1

(A)

In Problems 1 through 13 give (A) the union of the two sets and (B) the intersection of the two sets.

1. $[-3, 4]$, $[2, 7]$.

2. $[-10, -5]$, $[-6, -2]$.

3. $(-3, 4)$, $(2, 7)$.

4. $(-3, 4]$, $[2, 7)$.

5. $[-3, 4)$, $(2, 7]$.

6. $[-10, -5)$, $(-6, -2]$.

7. $(-10, -5)$, $[-6, -2)$.

8. $(-7, 5)$, $(1, 2)$.

9. $(-7, 5)$, $[1, 2]$.

10. $(-7, 5)$, $(1, 2]$.

11. $[-7, 5)$, $[1, 2)$.

12. $[1, 2]$, $[3, 5]$.

13. Let X be the set of all even integers and let Y be the set of all odd integers. find (A) $X \cup Y$, and (B) $X \cap Y$.

***14.** Guess at a good (reasonable) definition for the equality of two sets. In other words what should $X = Y$ mean?

***15.** Is an Associative Law true for $\cup$? In other words is it true that for all sets X, Y, and Z,

$$X \cup (Y \cup Z) = (X \cup Y) \cup Z?$$

***16.** Is an Associative Law true for $\cap$? In other words is it true that for all sets X, Y, and Z,

$$X \cap (Y \cap Z) = (X \cap Y) \cap Z$$

***17.** Is a Commutative Law true for (A) unions, (B) intersections?

18. Is the symbol $\Rightarrow$ transitive? This means is it true that:

$$\text{If } A \Rightarrow B, \quad \text{and} \quad B \Rightarrow C, \quad \text{then } A \Rightarrow C?$$

19. Is the symbol $\Leftrightarrow$ transitive?

20. Is the symbol $\subset$ commutative? Is it transitive?

---

2. The Elementary Theorems

As a starting point we assume that we understand the set of positive numbers, the operations of addition, subtraction, multiplication, division, and raising to a power (exponentiation). As we explained in Section 1, we use P to denote the set of all positive numbers. We use $P^{(-)}$ to denote the set of all negative numbers. We might prefer the letter N for the set of negative numbers, but this symbol N is reserved for the natural numbers (the counting numbers, or the positive integers). Also Q cannot be used for the negative numbers since it is reserved for the set of all rational numbers. The number zero occupies a special place in the set of real numbers. It is not in P nor in $P^{(-)}$ but is unique and forms a set $\{0\}$ all by itself. Thus the set of all real numbers R can be written as

$$R = P \cup \{0\} \cup P^{(-)}, \tag{1}$$

where no two of the sets have a point in common. (They are *pair wise disjoint.*)

We also assume the following fundamental properties that were proved in earlier chapters.

(I) A real number x is a negative number if and only if $-x$ is a positive number. In other words,

$$x \text{ is in } P^{(-)} \text{ if and only if } -x \text{ is in } P.$$

(II) The sum of any two positive numbers is a positive number. In other words,

$$\text{if } x \text{ and } y \text{ are in } P, \text{ then } x+y \text{ is in } P.$$

(III) The product of any two positive numbers is a positive number. In other words,

$$\text{if } x \text{ and } y \text{ are in } P, \text{ then } xy \text{ is in } P.$$

Mathematicians have a fancy phrase for (II) and (III). They say that a set is *closed under an operation*, if the operation always produces a member of the set from members of the set. Thus:

(II) states that P is closed under addition.
(III) states that P is closed under multiplication.

In our earlier work, an inequality $a < b$ was defined geometrically by the location of their corresponding points on the x-axis. Here we prefer to use an analytic definition of $a < b$.

Definition 1. $a < b$, (read "a is less than b") if and only if $b - a$ is positive. Under these conditions we also write $b > a$ (read "b is greater than a").

Clearly the analytic definition of $a < b$ and the geometric definition are *equivalent.* If $a < b$, under either definition, then this inequality also holds under the other definition.

From this definition we see that $0 < b$ if and only if b is a positive number (put $a = 0$ in the definition).

As trivial examples we have:

$$86 < 99 \quad \text{because} \quad 99 - 86 = 13 \text{ an element of } P.$$

$$-1000 < 1 \quad \text{because} \quad 1 - (-1000) = 1001 \in P.$$

$$\frac{21}{34} < \frac{55}{89} \quad \text{because} \quad \frac{55}{89} - \frac{21}{34} = \frac{1}{3026} \in P.$$

On the other hand it is not at all obvious that

$$\sqrt{83} + \sqrt{117} < \sqrt{99} + \sqrt{101},$$

and it will take some effort to prove this (see Problem 5 of Exercise 2).

Since $a < b$ and $b > a$ are equivalent, one of the two symbols $<$ and $>$ is unnecessary and could be dropped. However, it is convenient to have both of them available for use. It is also convenient to have a compound symbol $a \leq b$, which means that either $a < b$ or $a = b$. Similarly $a \geq b$ means that either $a > b$ or $a = b$.

Theorem 1. *If a and b are any two real numbers, then exactly one of three relations*

$$\text{(2)} \qquad \text{(A)} \quad a < b, \quad \text{(B)} \quad a = b, \quad \text{(C)} \quad b < a$$

holds (is true).

Proof. For any two real numbers either $b - a$ is positive, or $b - a = 0$, or $b - a$ is negative and these three cases are mutually exclusive. In the first case $a < b$, and in the second case $a = b$. In the third case $b - a$ is negative, so that $-(b - a) = a - b$ is positive. Hence $b < a$. ∎

Theorem 2. *If $a < b$ and c is any positive number, then*

$$ca < cb.$$

In other words a true inequality remains true when multiplied on both sides by the same positive number.

Proof. Since $a < b$, then $b - a$ is a positive number. Since the product of two positive numbers is again positive we have

$$c(b - a) = cb - ca$$

is positive. By Definition 1 this gives $ca < cb$. ∎

We leave for the student the proof of

Theorem 3. *If $a < b$ and c is any negative number, then*

$$ca > cb.$$

In other words, when an inequality is multiplied on both sides by the same negative number the inequality sign is reversed.

Theorem 4 (A Transitive Law). *If $a < b$ and $b < c$, then $a < c$.*

Proof. By hypothesis $b - a$ and $c - b$ are positive numbers. Then the sum $(b-a)+(c-b)$ is a positive number. But this sum is $c - a$. Since $c - a$ is positive, then by Definition 1, $a < c$. ∎

By a similar type of argument the student can prove

Theorem 5. *If $a < c$ and $b < d$, then $a + b < c + d$.*

In other words, two inequalities can be added term wise to give a true inequality. Of course, the inequality sign must be in the same direction in all three of the inequalities.

Theorem 6. *If $a < b$ and c is any number, then $a + c < b + c$.*

Proof. Clearly $(b + c) - (a + c) = b - a$ is positive. ∎

Thus a true inequality remains true when the same number is added to both sides. Note that c can be a negative number, so this theorem includes subtraction.

Theorem 7. *If $0 < a < b$, then*

$$\frac{1}{a} > \frac{1}{b} > 0\,. \tag{3}$$

Thus reciprocation reverses the inequality sign if both terms are positive.

Proof. Multiply both sides of the equation $a < b$ by $1/ab$ and use Theorem 2. ∎

Theorem 8. *If $0 < a < b$ and $0 < c < d$, then $ac < bd$.*

Thus the multiplication of the corresponding terms of two inequalities gives another inequality. Here all the terms must be positive, and the inequality sign must be in the same direction in all three inequalities.

Proof. If we use Theorem 2 the inequality $a < b$ will give $ac < bc$. Similarly, $c < d$ will give $bc < bd$. Then the transitive law, Theorem 4, will give $ac < bd$. ∎

Theorem 9. *If $0 < a < b$ and n is any positive integer, then*

$$a^n < b^n\,. \tag{4}$$

Proof. Apply Theorem 8, $n - 1$ times with $c = a$ and $d = b$. ∎

Theorem 10. *If $0 < a < b$ and n is any positive integer, then*

$$\sqrt[n]{a} < \sqrt[n]{b}\,,$$

where, if n is even $\sqrt[n]{\ }$ means the positive nth root.

Proof. The proof is a little complicated, because it uses the method of contradiction. By Theorem 1, there are only three possibilities:

$$\text{(A)}\quad \sqrt[n]{a} < \sqrt[n]{b}, \quad \text{(B)}\quad \sqrt[n]{a} = \sqrt[n]{b}, \quad \text{(C)}\quad \sqrt[n]{a} > \sqrt[n]{b}.$$

If we can prove (B) and (C) cannot be true, then it follows that (A) holds, and the proof is complete. In each of the last two cases we take the nth power of both sides. In case (B) we will find that $a = b$. But this is impossible because by hypothesis $a < b$. In case (C) we apply Theorem 9 to $\sqrt[n]{b} < \sqrt[n]{a}$ and find that $b < a$. Again this is impossible because $a < b$. Since (B) and (C) cannot be true the only case that can occur is (A). ■

In most of these theorems we can allow the equality sign to occur in the hypotheses as long as we make suitable changes in the conclusions. In this way we obtain a large number of new theorems that vary only slightly from the ones already proved. There is no need to list them all because they are really obvious. (The student who is in doubt on this point should investigate the matter on his/her own.) Merely as an illustration of the type of theorem to be expected, we give the following variation on Theorems 6 and 8.

Theorem 6′. *If $a \leq b$ and c is any number, then $a + c \leq b + c$.*

Theorem 8′. *If $0 < a \leq b$ and $0 < c < d$, then $ac < bd$.*

There is one other useful and very popular tool for proving inequalities, namely the innocent remark that the square of any real number c is either positive or zero, that is $c^2 \geq 0$.

Example 1. Prove that for any two real numbers a and b

$$2ab \leq a^2 + b^2\,, \tag{5}$$

and the equality sign holds if and only if $a = b$.

Solution. By the remark we just made

$$(a - b)^2 \geq 0\,, \tag{6}$$

with equality if and only if $a = b$. Therefore

$$a^2 - 2ab + b^2 \geq 0\,,$$

or

$$0 \leq a^2 - 2ab + b^2\,.$$

Then by Theorem 6′ (adding $2ab$ to both sides) $2ab \leq a^2 + b^2$. Since equality occurs in (6) if and only if $a = b$, it occurs in (5) under the same conditions. ►►

To check the inequality (5) we try $a = 7$ and $b = 8$. Then we should have $2ab = 2(7)(8) = 112 \leq 7^2 + 8^2 = 49 + 64 = 113$, and as predicted by (5), we have $112 < 113$. Thus we have a numerical check of this inequality. The curious student should test a few other values for a and b.

Example 2. Prove that if a, b, c, and d are any four positive numbers then

$$ab + cd \leq \sqrt{a^2+c^2}\,\sqrt{b^2+d^2}\,. \tag{7}$$

It is not easy to see the proper starting place for this problem, so we work backward. That is, we start with the inequality (7) and see if we can deduce one that we know to be true. This operation is called the *analysis* of the inequality.

Analysis. If (7) is true we can square both sides and obtain

$$a^2b^2 + 2abcd + c^2d^2 \leq (a^2+c^2)(b^2+d^2)\,, \tag{8}$$

or

$$a^2b^2 + 2abcd + c^2d^2 \leq a^2b^2 + c^2b^2 + a^2d^2 + c^2d^2\,. \tag{9}$$

Then on transposing (Theorem 6′) four terms cancel and we have

$$0 \leq c^2b^2 - 2abcd + a^2d^2 \tag{10}$$

or

$$0 \leq (cb-ad)^2\,. \tag{11}$$

But we know that this last inequality is true. Hence if we can reverse our steps we can prove that the given inequality is also true.

Solution. We begin with the known inequality (11) and on expanding we find that (10) is also true. Then adding

$$a^2b^2 + 2abcd + c^2d^2$$

to both sides of (10) will give (9). Factoring the right side of (9) will give (8). Finally, taking the square root of both sides of (8) gives (7) ▶▶

It is customary to do the analysis on scratch paper and then present the correct solution (as we have done here) by presenting the analysis in the reverse order.

The inequality (7) is a very special and very simple case of a far more complicated and deeper inequality known as the CBS inequality. We will meet the CBS inequality in Section 3 and prove it in Section 4.

Example 3. Without using tables or a calculator, prove that

$$\sqrt{2} + \sqrt{6} < \sqrt{3} + \sqrt{5}\,. \tag{12}$$

Solution. We give the analysis. If we square both sides of (12) we obtain

$$2 + 2\sqrt{2}\,\sqrt{6} + 6 < 3 + 2\sqrt{3}\,\sqrt{5} + 5\,, \tag{13}$$

or on subtracting 8 from both sides and dividing by 2 we have

$$\sqrt{12} < \sqrt{15}\,. \tag{14}$$

But since $12 < 15$, the inequality (14) is obviously true. To prove the given inequality (12), we start with the remark that $12 < 15$ and reverse the above steps. ▶▶

In practical work, we do not bother to reverse the steps of the above analysis. Knowing that this can be done, we do not bother to do it, unless this is an exam problem and the teacher demands that the student give the steps in the proper order. Going from (12) to (14) is sufficient for any mathematician (of course assuming that all of the steps are reversible).

We recall that the absolute value of x, denoted by $|x|$ is defined by

$$|x| = \begin{cases} x\,, & \text{if } x > 0\,, \\ 0\,, & \text{if } x = 0\,, \\ -x\,, & \text{if } x < 0\,. \end{cases} \tag{15}$$

The function $|x|$ has a number of very important properties that we state below in three theorems. However each proof is completed by examining all the possible cases for the variables: are they positive, zero, or negative. Thus the theorems are often called trivial by the polished mathematician. Examining cases is rather boring, and so we leave the proofs to the energetic (or skeptical) student.

Theorem 11. *For all real numbers x, we have $|x| \geq 0$, and*

$$-|x| \leq x \leq |x|\,. \tag{16}$$

Further $|x| = 0$ if and only if $x = 0$.

Theorem 12. *If x and y are any pair of real numbers, then*

$$|xy| = |x|\,|y|\,, \tag{17}$$

$$|x - y| = |y - x|\,, \tag{18}$$

$$\sqrt{x^2} = |x|\,, \tag{19}$$

and if $y \neq 0$, then

$$\left|\frac{x}{y}\right| = \frac{|x|}{|y|}\,. \tag{20}$$

Theorem 13 (The Triangle Inequality). *If x and y are any pair of real numbers, then*

$$|x + y| \leq |x| + |y|\,, \tag{21}$$

and

$$|x| - |y| \leq |x - y|\,. \tag{22}$$

Inequality (21) is called the *triangle inequality* because it resembles the fact that in any triangle the length of one side is less than or equal to the sum of the lengths of the other two sides. In Section 4 we will learn just why the inequality (21) resembles an inequality for the sides of a triangle.

Proof of Theorem 13. The inequality (21) becomes an equality when both x and y are positive, and is also an equality when both x and y are negative. If either $x = 0$ or $y = 0$, (or both) the inequality (21) is trivial. Suppose that x and y have opposite signs, then

$$|x + y| \leq \max\{|x|, |y|\} < |x| + |y|,$$

where max S means the largest number in the collection of numbers in S (there may be several such possible numbers, all equal). This concludes the proof of (21).

The proof of (22) is tricky. Now that we have proved (21) we can apply it to the *TWO* numbers y and $x - y$. Since the sum is x we find that

$$|x| = |y + (x - y)| \leq |y| + |x - y| \quad \text{or} \quad |x| \leq |y| + |x - y|.$$

This will give (22), when we add $-|y|$ to both sides. ■

Exercise 2

(A)

In Problems 1 through 4 determine which of the given numbers x or y is the greater and give a proof without using tables or a calculator.

1. $x = \sqrt{19} + \sqrt{21}, \quad y = \sqrt{17} + \sqrt{23}.$

2. $x = 4 + \sqrt{15} \quad y = \sqrt{17} + \sqrt{14}.$

3. $x = \sqrt{17} + 4\sqrt{5}, \quad y = 5\sqrt{7}.$

4. $x = 2\sqrt{2}, \quad y = \sqrt[3]{23}.$

5. Prove that if $1 < B < A$, then

$$\sqrt{A - B} + \sqrt{A + B} < \sqrt{A - 1} + \sqrt{A + 1}.$$

Notice that this problem contains Problem 1 as a special case.

6. Prove that the inequality of Example 2, Equation (7), is true even if some or all of the variables are negative.

In Problems 7 through 18 prove the given inequality under the assumption that all the quantities involved are positive. In each problem determine the conditions under which the equality sign occurs.

7. $a + 1/a \geq 2.$

8. $\dfrac{a}{5b} + \dfrac{5b}{4a} \geq 1.$

9. $\sqrt{\dfrac{c}{d}} + \sqrt{\dfrac{d}{c}} \geq 2$

10. $(c + d)^2 \geq 4cd.$

11. $\dfrac{a + b}{2} \geq \sqrt{ab} \geq \dfrac{2ab}{a + b}.$

12. $(a + 5b)(a + 2b) \geq 9b(a + b).$

13. $x^2 + 4y^2 \geq 4xy$.

14. $x^2 + y^2 + z^2 \geq xy + yz + zx$.

15. $\dfrac{c^2}{d^2} + \dfrac{d^2}{c^2} + 6 \geq \dfrac{4c}{d} + \dfrac{4d}{c}$.

16. $\dfrac{a+3b}{3b} \geq \dfrac{4a}{a+3b}$.

17. $cd(c+d) \leq c^3 + d^3$.

18. $4ABCD \leq (AB+CD)(AC+BD)$.

19. Which of the inequalities in Problems 7 through 18 are meaningful and still true if some or all of the letters may represent negative numbers.

20. Which of the following relations are true for all real numbers x, y, and z?

A. $|xyz| = |x|\,|y|z|$.

B. $|x+y+z| \leq |x| + |y| + |z|$.

C. $|x-y| \leq |x-z| + |y-z|$.

***21.** We recall from Chapter 4, Section 4 that for each positive integer n the symbol $n!$ (read "n factorial") represents the product of all the integers between 1 and n inclusive. For example

$$5! = 1(2)(3)(4)(5) = 120\,.$$

To complete the definition we put $0! = 1$. Prove that if $n \geq 1$, then $n^n \leq (n!)^2$.

22. Let p/q be any positive rational number, prove that if $0 < a < b$, then

$$a^{p/q} < b^{p/q}\,.$$

3. Euclidean n-Dimensional Space

To understand some of the inequalities in the next section we must look (for a short time) at n-dimensional Euclidean space denoted by $E^{(n)}$. Please don't be scared by these fancy words. We do not "look at" $E^{(n)}$ because I do not believe that anyone can really see $E^{(n)}$ if $n > 3$. But it is very easy to understand $E^{(n)}$ as we will now explain. In the plane, the coordinates of a point have the form (x_1, x_2) where x_1 and x_2 are any pair of real numbers. (Notice that we used x_2 in the second place rather than y. This may be upsetting to the beginner, but it is perfectly legal and the reason for this change will appear soon.) Thus $E^{(2)}$ can be regarded as the set of all possible pairs of real numbers. Similarly, $E^{(3)}$, the usual 3-dimensional space, is the set of all triples (x_1, x_2, x_3) of real numbers. Of course, before we can accept the collection of all pairs (x_1, x_2) or all triples (x_1, x_2, x_3) as the space $E^{(2)}$, or $E^{(3)}$ certain rules of combination and operation must hold.

It is now a simple matter to define n-dimensional Euclidean space $E^{(n)}$ as the set of all n-tuples $(x_1, x_2, x_3, \ldots, x_n)$ of real numbers, called *points* of $E^{(n)}$. For example $(3, 1, 5, 6, 4, 2)$

and $(2,2,4,3,3,3)$ are both points in $E^{(6)}$, 6-dimensional Euclidean space. Each x_k in $P(x_1, x_2, x_3, \ldots, x_n)$ is called a *coordinate* of the point P.

If $X(x_1, x_2, x_3, \ldots, x_n)$ and $Y(y_1, y_2, y_3, \ldots, y_n)$ are any two points in $E^{(n)}$, then we denote the distance from X to Y by $|XY|$ and this distance is *defined* by

$$|XY| = \sqrt{(y_1 - x_1)^2 + (y_2 - x_2)^2 + (y_3 - x_3)^2 + \cdots + (y_n - x_n)^2} = \sqrt{\sum_{k=1}^{n} (y_k - x_k)^2}\,. \tag{23}$$

If the use of the sum sign $\sum$, under the radical bothers you it may be worthwhile to review the explanation of $\sum$ given in Chapter 4, Section 2. We will see $\sum$ frequently in this chapter.

Thus, if X and Y are $(3,1,5,6,4,2)$ and $(2,2,4,3,3,3)$ from the previous paragraph, then the distance from X to Y is

$$|XY| = \sqrt{(2-3)^2 + (2-1)^2 + (4-5)^2 + (3-6)^2 + (3-4)^2 + (3-2)^2} = \sqrt{14}\,.$$

If the point Y is the *origin*, $O(0,0,0,\ldots,0)$, in $E^{(n)}$, then according to Equation (23) the distance of X from the origin is

$$|OX| = \sqrt{x_1^2 + x_2^2 + x_3^2 + \cdots + x_n^2}\,, \tag{24}$$

obtained by setting $y_k = 0$ in (23) for each k.

We also define a *translation of* $E^{(n)}$ by $T(t_1, t_2, \ldots, t_n)$ as the set of points obtained when each point $P(x_1, x_2, x_3, \ldots, x_n)$ is replaced by $P'(x_1 + t_1, x_2 + t_2, x_3 + t_3, \ldots, x_n + t_n)$. This is a *rigid motion*: (The distance between any two points remains unchanged by a translation). We can use a suitable translation to move any preassigned point to the origin. If we want to examine the angle at C of a triangle ACB in $E^{(n)}$ we first translate the point C to the origin (see Figure 2 for the plane case).

Before we continue with our study of $E^{(n)}$ we must look at some facts about $E^{(2)}$, the plane case. Here we must use a little trigonometry. The student who is not familiar with

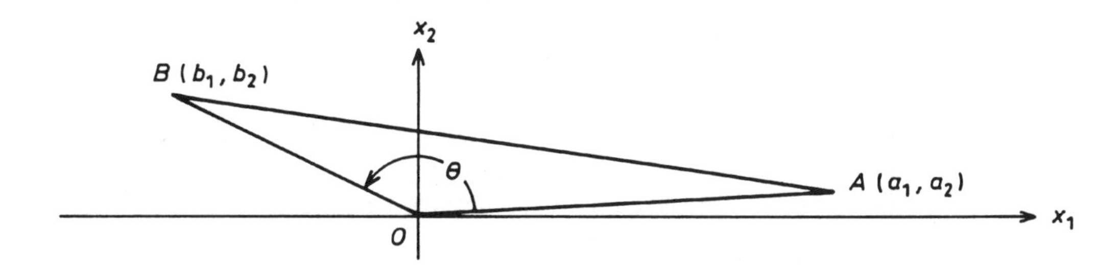

Figure 2.

trigonometry will find the material he/she needs on the Cosine Law, in Chapter 22, Section 5 but we will need the Cosine Law for further work.

For the triangle AOB shown in Figure 2 the Cosine Law gives

$$|AB|^2 = |OA|^2 + |OB|^2 - 2|OA|\,|OB|\cos\theta\,, \tag{25}$$

where θ (read "theta") is the angle AOB at the vertex O of the triangle. We find the lengths of the sides of the triangle from Figure 2 and use these in Equation (25). Special cases ($n = 2$) of Equations (23) and (24) help us to write (25) as

$$(b_1 - a_1)^2 + (b_2 - a_2)^2 = (a_1^2 + a_2^2) + (b_1^2 + b_2^2) - 2|OA|\,|OB|\cos\theta\,. \tag{26}$$

When we expand the left side, four of the square terms cancel with four of the terms on the right side (the student should check this) and all that is left is

$$-2a_1b_1 - 2a_2b_2 = -2|OA|\,|OB|\cos\theta\,,$$

or

$$\boxed{a_1b_1 + a_2b_2 = |OA|\,|OB|\cos\theta\,.} \tag{27}$$

Equation (27) gives a meaning to the expression on the left side, and this is the meaning we were searching for. If we divide by the lengths of the sides on the right side of (27) we arrive at

$$\boxed{\frac{a_1b_1 + a_2b_2}{\sqrt{a_1^2 + a_2^2}\,\sqrt{b_1^2 + b_2^2}} = \cos\theta\,.} \tag{28}$$

Now it is well-known that $-1 \le \cos\theta \le 1$, and since the denominator in (28) is always positive, multiplication will give the double inequality

$$\boxed{-\sqrt{a_1^2 + a_2^2}\,\sqrt{b_1^2 + b_2^2} \le a_1b_1 + a_2b_2 \le \sqrt{a_1^2 + a_2^2}\,\sqrt{b_1^2 + b_2^2}\,.} \tag{29}$$

The same proof will give the same type of result in 3-dimensional space. If we use summation notation this result is

$$\boxed{\frac{\sum_{k=1}^{3} a_kb_k}{\sqrt{\sum_{k=1}^{3} a_k^2}\,\sqrt{\sum_{k=1}^{3} b_k^2}} = \cos\theta\,,} \qquad \text{(compare this with (28))}\,, \tag{30}$$

and the same double inequality

$$(31)\qquad \boxed{-\sqrt{\sum_{k=1}^{3} a_k^2}\sqrt{\sum_{k=1}^{3} b_k^2} \le \sum_{k=1}^{3} a_k b_k \le \sqrt{\sum_{k=1}^{3} a_k^2}\sqrt{\sum_{k=1}^{3} b_k^2}\,.}$$

What happens to the relations (30) and (31) when we change the range of the summation from 1, 2, 3 to 1, 2, 3, ..., n, that is, when we go over to n-dimensional Euclidean space. Answer: (31) remains a true inequality which we display as

$$(32)\qquad \boxed{\begin{aligned} &-\sqrt{a_1^2+a_2^2+a_3^2+\cdots+a_n^2}\,\sqrt{b_1^2+b_2^2+b_3^2+\cdots+b_n^2} \\ &\le a_1b_1+a_2b_2+a_3b_3+\cdots+a_nb_n \\ &\le \sqrt{a_1^2+a_2^2+a_3^2+\cdots+a_n^2}\,\sqrt{b_1^2+b_2^2+b_3^2+\cdots+b_n^2} \end{aligned}}$$

and which still needs a proof. This beautiful inequality is called the CBS inequality. (The letters CBS will be explained and the proof of (32) will be given in the next section.)

What happens to Equation (30) when we go over to $E^{(n)}$ when $n > 3$? Answer: It becomes a definition, the definition of an angle in n-dimensional space for $n > 3$.

Definition 2 (Angle). If θ is the angle at O of the triangle AOB where A is the point $A(a_1, a_2, a_3, \ldots, a_n)$ and B is the point $B(b_1, b_2, b_3, \ldots, b_n)$ in $E^{(n)}$, then

$$(33)\qquad \boxed{\cos\theta = \frac{a_1b_1+a_2b_2+a_3b_3+\cdots+a_nb_n}{\sqrt{a_1^2+a_2^2+a_3^2+\cdots+a_n^2}\,\sqrt{b_1^2+b_2^2+b_3^2+\cdots+b_n^2}}\,.}$$

Is this definition meaningful? Is there a real angle determined by Equation (33)? The answer is yes, if we restrict θ to lie in the interval $[0, \pi]$, and if $|\cos\theta| \le 1$. But this last condition is always satisfied because this is just the statement of the CBS inequality (32), and (as we have already mentioned) this inequality will be proved in the next section.

Once more we leave $E^{(n)}$ for the plane. From Figure 2 it is obvious that for any triangle AOB in the plane, the length of any one side is less than or equal to the sum of the lengths of the other two sides. This gives the inequality

$$(34)\qquad \boxed{|AB| \le |OA| + |OB|\,.}$$

This inequality is called the *triangle inequality* for obvious reasons. The very same inequality is true in n-dimensional Euclidean space. In $E^{(n)}$ the inequality (34) has the form

$$(35)\qquad \boxed{\begin{aligned} &\sqrt{(b_1-a_1)^2+(b_2-a_2)^2+(b_3-a_3)^2+\cdots+(b_n-a_n)^2} \\ &\le \sqrt{a_1^2+a_2^2+a_3^2+\cdots+a_n^2}+\sqrt{b_1^2+b_2^2+b_3^2+\cdots+b_n^2}\,. \end{aligned}}$$

If the inequality (35) is true in $E^{(n)}$ we must prove it.

Theorem 14 (The Triangle Inequality). *For any two points A and B in $E^{(n)}$ the inequality* (35) *is true.*

The proof of Theorem 14 will also be given in the next section.

Exercise 3

(A)

In Problems 1 through 6 find the distance of the given point from the origin. All points are in $E^{(n)}$, with $n = 4, 5$ or 6.

1. $A(1,1,1,1)$.

2. $B(1,-1,-1,1)$.

3. $C(1,2,3,4)$.

4. $D(5,4,3,2)$.

5. $E(3,-4,3,4-4)$.

6. $F(1,-2,-3,-4,6)$.

In Problems 7 through 12 find the distance from A to B for the given pair of points.

7. $A(1,1,1,1)\,,\ B(1,-1,-1,1)$.

8. $A(6,5,4,3)\,,\ B(5,3,1,-1)$.

9. $A(1,2,3,4)\,,\ B(5,4,3,2)$.

10. $A(3,-4,3,4{-}4)\,,\ B(1,-2,-3,-4,6)$.

11. $A(1,3,5,7,9)\,,\ B(2,4,6,8,9)$.

12. $A(1,-1,1,-1,1,2)\,,\ B(2,0,2,0,3,4)$.

In Problems 13 through 18 find $\cos\theta$ where θ is the angle at O for the triangle AOB where A and B are the points given in Problems 7 through 12.

13. Problem 7.

14. Problem 8.

15. Problem 9.

16. Problem 10.

17. Problem 11.

18. Problem 12.

19. Is $|\cos\theta| \leq 1$ for the angle θ in Problems 13 through 18?

20 Let $P = P(a,b,c)$, where each one of a, b, and c is either 0 or 1. These eight points (count them) are the *vertices* of a unit cube in the first octant of $E^{(3)}$. How many vertices does a similar unit cube have: (I) in $E^{(4)}$, (II) in $E^{(5)}$, (III) in $E^{(n)}$? P is a vertex if each coordinate is either 0 or 1.

21. Given two fixed points X and Y in $E^{(n)}$, is there a translation of $E^{(n)}$ that will take X into Y. If so find it.

22. Prove that if $E^{(n)}$ is translated by $T(t_1, t_2, \ldots, t_n)$, then for any two points the distance between them is not changed by this translation.

23. How would you define "adjacent vertices" in $E^{(n)}$?

24. How would you define a "diagonal" of a unit cube in $E^{(n)}$?

25. Find the length of the longest diagonal for the unit cube in $E^{(3)}$ described in Problem 20.

26. Find the length of the longest diagonal for the unit cubes described in Problem 20 for: (I) $E^{(4)}$, (II) $E^{(5)}$, (III) $E^{(n)}$.

---

4. Some Famous Inequalities

Some inequalities are so important for research in advanced mathematics that they are given special titles, together with a royal treatment. In this section we prove a few of these.

Let

$$P = (a_1, a_2, a_3, \ldots, a_n)$$

and

$$Q = (b_1, b_2, b_3, \ldots, b_n)$$

be two points in $E^{(n)}$. In other words, P and Q are two sequences of real numbers.

Theorem 15. *If P, and Q are any two points in $E^{(n)}$ (or any two sequences of n real numbers), then (see Equation (32) of Section 3)*

$$a_1b_1 + a_2b_2 + a_3b_3 + \cdots + a_nb_n \le \sqrt{a_1^2 + a_2^2 + a_3^2 + \cdots + a_n^2}\sqrt{b_1^2 + b_2^2 + \cdots + b_n^2}\,. \tag{36}$$

Equality occurs in (36) if and only if there is a constant t such that $b_k/a_k = t$ for each $k = 1, 2, 3, \ldots, n$.

This theorem is known as the Cauchy–Bunyiakowski–Schwarz inequality (after the three mathematicians who found it independently). Since the list contains one Frenchman, one Russian, and one German, various subsets of these names are used in accordance with the nationality of the writer. The inequality (36) is also known as the CBS inequality.

The inequality (36) can be written more compactly if we are willing to use summation notation (described in detail in Chapter 4.) Briefly, (see equation (31) of Section 3) the inequality (36) becomes

$$\sum_{k=1}^{n} a_kb_k \le \sqrt{\sum_{k=1}^{n} a_k^2}\sqrt{\sum_{k=1}^{n} b_k^2}\,. \tag{37}$$

Don't let Equation (37) scare you. It says exactly the same thing that (36) says.

How about the proof of Theorem 15? It is one of the prettiest little pieces of mathematics that I know of, and I recommend that the student memorize it strictly for his/her pleasure.

Proof of Theorem 15. For each $k = 1, 2, 3, \ldots, n$ and any real x we have

$$(a_k x - b_k)^2 \geq 0\,, \tag{38}$$

or

$$a_k^2 x^2 - 2a_k b_k x + b_k^2 \geq 0\,. \tag{39}$$

We now add all of the inequalities (39) for $k = 1, 2, 3, \ldots, n$ and on factoring the common terms x^2 and x we obtain

$$Ax^2 - Bx + C \geq 0\,, \tag{40}$$

where

$$A = \sum_{k=1}^{n} a_k^2\,, \quad B = 2\sum_{k=1}^{n} a_k b_k\,, \quad \text{and} \quad C = \sum_{k=1}^{n} b_k^2\,. \tag{41}$$

For all real x each of the expressions (38) or (39) is positive or zero so the sum has the same property. Further the sum (40) is positive unless all of the expressions (38) or (39) are zero for the SAME value of x (which we denote by t). Thus, the quadratic equation $Ax^2 - Bx + C = 0$ either has no real solutions, or a double root at $x = t$. Therefore the discriminant $B^2 - 4AC \leq 0$. If now, we use the expressions (41) for A, B, and C, the inequality $B^2 - 4AC \leq 0$ becomes

$$\left(2\sum_{k=1}^{n} a_k b_k\right)^2 - 4\left(\sum_{k=1}^{n} a_k^2\right)\left(\sum_{k=1}^{n} b_k^2\right) \leq 0\,. \tag{42}$$

When we cancel 4, put the product of the last two sums on the right side, and take the square root of both sides we obtain (36), the inequality of the theorem.

Equality in (36) occurs if and only if all the terms in (38) are zero simultaneously, that is, if and only if there is a t such that $a_k t - b_k = 0$ for every k. In this case $b_k/a_k = t$ for every k for which $a_k \neq 0$. If for some special k we have $a_k = 0$, then the corresponding $b_k = 0$ and the points are in $E^{(n-1)}$. ∎

If equality occurs in (36), then all of the points (a_k, b_k) lie on a fixed line through the origin.

The importance of Theorem 15 for geometry in $E^{(n)}$ has been explained earlier in Section 3.

Example 1. Check Theorem 15, using the two points

$$P = (1, 2, 3, 4)\,, \quad \text{and} \quad Q = (2, 3, 4, 5)\,.$$

Solution. With these two points the left side of (36) is

$$L = 1(2) + 2(3) + 3(4) + 4(5) = 2 + 6 + 12 + 20 = 40\,.$$

The right side is $R = \sqrt{1+4+9+16}\,\sqrt{4+9+16+25} = \sqrt{30(54)}$, and this is approximately 40.2492, so the assertion $L < R$ holds, just as the theorem predicted. ▶▶

Our next inequality concerns the relationship among three averages or *means* (as they are properly called). These means are the *arithmetic mean*, the *geometric mean*, and the *harmonic mean*. Given a sequence of POSITIVE numbers $B^* = (b_1, b_2, b_3, \ldots, b_n)$, these means are given the symbols $A(B^*)$, $G(B^*)$, and $H(B^*)$ and are defined by

$$\text{(43)} \qquad \text{Arithmetic Mean} \quad A(B^*) = \frac{b_1 + b_2 + b_3 + \cdots + b_n}{n},$$

$$\text{(44)} \qquad \text{Geometric Mean} \quad G(B^*) = \sqrt[n]{b_1 b_2 b_3 \cdots b_n}\,,$$

$$\text{(45)} \qquad \text{Harmonic Mean} \quad H(B^*) = \frac{n}{\dfrac{1}{b_1} + \dfrac{1}{b_2} + \dfrac{1}{b_3} + \cdots + \dfrac{1}{b_n}}\,.$$

For brevity we denote these means by A, G, and H.

Theorem 16. *For any sequence of positive numbers B^*,*

$$\text{(46)} \qquad \boxed{H \le G \le A\,,}$$

with equality if and only if all b_k are the same.

We observe that the inequality (46) is easy to memorize because the terms are in anti-alphabetical order.

Proof. Suppose first that all b_k are the same number X. Then it is obvious that $H = G = A = X$ (the industrious student should check this).

Next we concentrate on the inequality $G \le A$. If M is the maximum b_k in the sequence B^*, then $G \le M$ and $A \le M$. Let m and M be the smallest and largest numbers in the sequence B^* and assume that $m < M$. We will replace these two numbers by two new numbers. The new numbers will be selected so that G increases, but A stays the same. Suppose that $d = (M-m)/2$. (It might help the student to make a little diagram at this point.) We replace m and M by $m + d$ and $M - d$ respectively. Thus if $m = b_1$ and $M = b_2$, then the new sequence is $B_1^* = (m+d, M-d, b_3, b_4, \ldots, b_n)$. The arithmetic mean for B_1^* is the same as for B^* because $(m+d) + (M-d) = m + M = b_1 + b_2$. On the other hand the geometric mean has increased because

$$(m+d)(M-d) = mM + d(M-m) - d^2 = mM + d(2d) - d^2 = mM + d^2$$

and this is larger than $mM = b_1 b_2$. Thus, in progressing from the sequence B^* to B_1^*, the geometric mean has increased, but the arithmetic mean has stayed the same. But we removed the minimum and maximum members (m and M) from the sequence B^*. We can repeat this process. In the next step we replace the minimum and maximum members of the new sequence B_1^*, by their average. Thus, little by little the original sequence gets closer and closer to a sequence in which all the numbers are the same. At each stage G has increased and A remains unchanged. Now when all the elements are the same we have $G = A$. Hence at the beginning we must have $G < A$. This completes the proof of the second inequality in (46).

To prove that $H \leq G$, we apply the inequality $G \leq A$ to the sequence obtained by replacing each b_k by its reciprocal $1/b_k$. Then for this new sequence, the inequality $G \leq A$ gives

$$\sqrt[n]{\frac{1}{b_1}\frac{1}{b_2}\frac{1}{b_3}\cdots\frac{1}{b_n}} \leq \frac{\frac{1}{b_1}+\frac{1}{b_2}+\frac{1}{b_3}+\cdots+\frac{1}{b_n}}{n}. \tag{47}$$

Now take the reciprocals of both sides of (47) and use Theorem 7. This will give $G \geq H$, or its equivalent $H \leq G$. ■

Example 2. Check Theorem 16 using the sequence $(1, 2, 4, 5)$

Solution. For the sequence $(1, 2, 4, 5)$,

$$\begin{aligned} H &= \frac{4}{\frac{1}{1}+\frac{1}{2}+\frac{1}{4}+\frac{1}{5}} = \frac{4}{1+0.5+0.25+0.2} = \frac{4}{1.95} \\ &\approx 2.0513 \text{ (approximately)}. \\ G &= \sqrt[4]{1(2)(4)(5)} = \sqrt[4]{40} \approx 2.5149 \text{ (approximately)}. \\ A &= (1+2+4+5)/4 = 3. \quad \text{As expected, } H < G < A. \end{aligned}$$

►►

Before we take up the next inequality we need some modest preparation. First we review Theorem 15, and look again at the inequality

$$\begin{aligned} &a_1b_1 + a_2b_2 + a_3b_3 + \cdots + a_nb_n \\ &\leq \sqrt{a_1^2 + a_2^2 + a_3^2 + \cdots + a_n^2}\sqrt{b_1^2 + b_2^2 + b_3^2 + \cdots + b_n^2}. \end{aligned} \tag{36}$$

Probably, while proving (36), the reader had in mind that all the terms were positive, but (36) is true even if many of the terms are negative. If we replace each negative term by its absolute value the right side is unchanged. Therefore (going over to $\sum$ notation) the CBS inequality gives the double inequality

$$\boxed{-\sqrt{\sum_{k=1}^{n} a_k^2}\sqrt{\sum_{k=1}^{n} b_k^2} \leq \sum_{k=1}^{n} a_kb_k \leq \sqrt{\sum_{k=1}^{n} a_k^2}\sqrt{\sum_{k=1}^{n} b_k^2},} \tag{48}$$

where we have added a negative lower bound in (48).

We recall from Section 3 that we proved that if AOB is any triangle in the plane with vertex at the origin and θ is the angle at that vertex then

$$\cos\theta = \frac{a_1b_1 + a_2b_2}{\sqrt{a_1^2 + a_2^2}\sqrt{b_1^2 + b_2^2}}, \tag{49}$$

(see Equation (28) of Section 3).

We also recall from Section 3, that in $E^{(n)}$ we have

Definition 2 (Angle). If θ is the angle at O of the triangle AOB where A is the point $A(a_1, a_2, a_3, \ldots, a_n)$ and B is the point $B(b_1, b_2, b_3, \ldots, b_n)$, then (by definition)

$$\cos\theta = \boxed{\frac{a_1b_1 + a_2b_2 + a_3b_3 + \cdots + a_nb_n}{\sqrt{a_1^2 + a_2^2 + a_3^2 + \cdots + a_n^2}\sqrt{b_1^2 + b_2^2 + b_3^2 + \cdots + b_n^2}}}. \tag{50}$$

We can now understand one of the many important applications of the CBS inequality. Here in (50) the CBS inequality shows that the definition of θ is meaningful because $|\cos\theta| \leq 1$.

From Figure 2 it is obvious that for any triangle AOB in the plane we have

$$\boxed{|AB| \leq |OA| + |OB|,} \tag{51}$$

(see Equation (34) of Section 3).

This inequality is called the *triangle inequality* for obvious reasons. The very same inequality is true in n-dimensional Euclidean space. This is the content of

Theorem 17 (The triangle inequality). *For any two points A and B in $E^{(n)}$ the inequality* (51) *is true.*

When the distances are used, (51) is equivalent to

$$\boxed{\begin{aligned}&\sqrt{(b_1 - a_1)^2 + (b_2 - a_2)^2 + (b_3 - a_3)^2 + \cdots + (b_n - a_n)^2} \\ &\quad\leq \sqrt{a_1^2 + a_2^2 + a_3^2 + \cdots + a_n^2} + \sqrt{b_1^2 + b_2^2 + b_3^2 + \cdots + b_n^2}.\end{aligned}} \tag{52}$$

Proof. We are to prove that for $A(a_1, a_2, a_3, \ldots, a_n)$ and $B(b_1, b_2, b_3, \ldots, b_n)$ the inequality (52) is always true (see Equation (35) of Section 3).

We can start from (52) and give an analysis. Having done this analysis, we will now reverse the order of events and proceed to prove (52) directly. The industrious reader might want to do his/her own analysis first.

We multiply all three terms in the CBS inequality (48) by -2. When the inequality signs are reversed and the terms written in the reverse order, this will give

$$-2\sqrt{\sum_{k=1}^{n} a_k^2}\sqrt{\sum_{k=1}^{n} b_k^2} \leq -2\sum a_kb_k \leq 2\sqrt{\sum_{k=1}^{n} a_k^2}\sqrt{\sum_{k=1}^{n} b_k^2}. \tag{53}$$

Now add to each of the three expressions in (53) the sums

$$\sum_{k=1}^{n} a_k^2 + \sum_{k=1}^{n} b_k^2 .$$

Then all three sums in (53) become perfect squares:

$$(54) \qquad \left(\sqrt{\sum_{k=1}^{n} a_k^2} - \sqrt{\sum_{k=1}^{n} b_k^2}\right)^2 \le \sum_{k=1}^{n} (a_k - b_k)^2 \le \left(\sqrt{\sum_{k=1}^{n} a_k^2} + \sqrt{\sum_{k=1}^{n} b_k^2}\right)^2 .$$

Since all the quantities are positive we may take the square roots of all three terms in (54). The last two terms will give (52). The first two terms in (54) will give $|OA| - |OB| \le |AB|$. ∎

Exercise 4

(A)

In Problems 1 through 6 check numerically that $H \le G \le A$ for the given sequence. You may use a calculator and approximate the value of the mean to three decimal places.

1. $A(7, 7, 7, 7)$. **2.** $B(1, 1, 2, 2)$.

3. $C(1, 3, 5, 8)$. **4.** $D(5, 4, 2, 2)$.

5. $E(3, 4, 3, 4, 3)$. **6.** $F(1, 2, 5, 8, 10)$.

7. Explain why we insist that all elements of the sequence are positive for the inequality $H \le G \le A$.

In Problems 8 through 13 check that the triangle inequality $|AB| \le |OA| + |OB|$ is satisfied for the given points A and B.

8. $A(1, 1, 1, 1)$, $B(1, -1, -1, 1)$. **9.** $A(6, 5, 4, 3)$, $B(5, 3, 1, -1)$.

10. $A(1, 2, 3, 4)$, $B(5, 4, 3, 2)$. **11.** $A(3, -4, 3, 4 - 4)$, $B(1, -2, -3, -4, 6)$.

12. $A(1, 3, 5, 7, 9)$, $B(2, 4, 6, 8, 9,)$. **13.** $A(1, -1, 1, -1, 1, 2)$, $B(2, 0, 2, 0, 3, 4)$.

***14.** Under what condition does the equality sign hold in the triangle inequality?

In Problems 15 through 20 check that the CBS inequality is correct by computing L and R and showing that $L \le R$ where

$$L = a_1 b_1 + a_2 b_2 + a_3 b_3 + \cdots + a_n b_n ,$$

and

$$R = \sqrt{a_1^2 + a_2^2 + a_3^2 + \cdots + a_n^2}\,\sqrt{b_1^2 + b_2^2 + b_3^2 + \cdots + b_n^2}\,.$$

15. $A(1,1,1,1)\,,\ B(2,2,2,2)$.

16. $A(6,5,4,3)\,,\ B(5,4,3,2)$.

17. $A(1,2,3,4)\,,\ B(3,4,5,6)$.

18. $A(3,-4,3,-4,3)\,,\ B(4,-3,4,-3,4)$.

19. $A(1,3,5,7,9)\,,\ B(2,4,6,8,10)$.

20. $A(3,2,1,-1,-2,-3)\,,$
$B(4,3,2,-1,-3,-4)$.

***21.** Try to reproduce the proof of the CBS inequality without looking at the book.

****22.** Do Problem 21 for the $H \le G \le A$ inequality.

****23.** Does the proof given in Section 4 that $G \le A$ contain an error? This is part of Equation (46) in Theorem 16.

---

5. Conditional Inequalities

Some inequalities are true only when the variables involved are restricted to lie in a certain domain. Such inequalities are called *conditional inequalities.*

Example 1. Solve the inequality $5x + 3 > 2x - 18$.

This means find the domain (values of x) for which the inequality is true.

Solution. Clearly, this inequality is equivalent to $3x > -21$. Hence the solution is the *ray* $x > -7$. ▶▶

Example 2. Solve the inequality

$$2x^2 - 3 < x^2 + 4x + 18\,.$$

Solution. Clearly this inequality is equivalent to

$$x^2 - 4x - 21 < 0\,,$$

$$(x-7)(x+3) < 0\,. \tag{55}$$

If $|x|$ is sufficiently large, then either both factors in (55) are positive, or both factors are negative. In either case the product is positive and the inequality is false. Thus we only need to know where there is a change of sign in the product. Let us imagine that x is a variable that runs from a negative number with large absolute value to a large positive number (the point x runs continuously from left to right). We may express this idea by saying that x varies from $-\infty$ to $+\infty$. The product in (55) changes only when x arrives at a zero of the product. In this problem the change in sign can occur only at $x = -3$ and $x = 7$. These points divide the x-axis into three sets (intervals or rays) namely $(-\infty, -3)$, $(-3, 7)$ and $(7, \infty)$. We can determine the sign of the product in any of these sets by computing the value at a single point of the set. Here, we select $x = 0$, and find that the product is $(-7)(3) < 0$. Consequently the inequality (55) is true for the entire interval $(-3, 7)$ and for no other value of x. Thus the solution to the inequality is the open interval $(-3, 7)$. ▶▶

Example 3. Solve the inequality

$$\frac{x-1}{x-3} > \frac{x+2}{x+4}. \tag{56}$$

Solution. There is a great temptation to "cross multiply" and get

$$(x-1)(x+4) > (x-3)(x+2). \tag{57}$$

This move is WRONG. We do not know if the multipliers are positive or negative so (57) may not be equivalent to (56). The correct move is to subtract the right side of (56) from both sides of the inequality. This gives

$$\frac{x-1}{x-3} - \frac{x+2}{x+4} > 0,$$

$$\frac{(x-1)(x+4)-(x-3)(x+2)}{(x-3)(x+4)} = \frac{x^2+3x-4-(x^2-x-6)}{(x-3)(x+4)} > 0,$$

or

$$\frac{4x+2}{(x-3)(x+4)} > 0. \tag{58}$$

As in Example 2, (as x increases from $-\infty$ to ∞) the left side changes sign at $x = -4$, $x = -1/2$, and $x = 3$ and only at these points. Computing the left side of (58) for $x = -5$, $x = -2$, $x = 1$, and $x = 4$, we see that the left side of (58) is positive if and only if $-4 < x < -1/2$ or $3 < x$. The solution can be stated as the set $(-4, -1/2) \cup (3, \infty)$. ▶▶

Example 4. Solve the inequality

$$Q = \frac{(x-1)(x-3)(x-5)(x-7)}{(x-2)(x-4)(x-6)(x-8)} < 0. \tag{59}$$

Solution. It is clear that the left side changes sign if x is any one of the integers 1, 2, 3, 4, 5, 6, 7, and 8. If $|x|$ is large, then $Q > 0$ and the inequality is false. One would then guess that the solution is the collection of open intervals

$$(1,2) \cup (3,4) \cup (5,6) \cup (7,8).$$

A careful analysis (as was done in Examples 2 and 3) will confirm this guess. ▶▶

Example 5. Solve the inequality $|15 - 4x| \leq 7$.

Solution. The given inequality is true if and only if

$$-7 \leq 15 - 4x \leq 7. \tag{60}$$

This gives two linear inequalities, both of which must be true. The left side of (60) gives $4x \leq 15 + 7 = 22$, or $x \leq 11/2$. The right side of (60) gives $-4x \leq 7 - 15 = -8$, or $x \geq 2$. Thus the solution is the closed interval [2,11/2]. ▶▶

Exercise 5

(A)

In Problems 1 through 29 solve the given inequality.

1. $2x - 9 > 13$.
2. $3x - 7 < 35$.
3. $4x + 5 \geq -12$.
4. $3x + 2 \leq 4$.
5. $x + 2 \leq 3x - 4$.
6. $7x - 8 < 9x + 10$.
7. $7x + 9 > 11$.
8. $\dfrac{4 - 5x}{2} < -13$.
9. $\dfrac{-3x + 7}{4} \geq -13$.
10. $\dfrac{6x - 2}{3x + 4} > 2$.
11. $\dfrac{5x - 4}{3x + 2} < 1$.
12. $\dfrac{8x - 5}{5x - 8} \geq 2$.
13. $x^2 - 10 \geq 0$.
14. $4x^2 + 4x - 15 < 0$.
***15.** $x^2 - 4x + 6 < 0$.
16. $13x^2 + 9x - 7 \leq x^2 - 3x + 5$.
17. $2x^2 + 3x - 4 \geq x^2 + 2x + 2$.
18. $|x - 5| < 7$.
19. $|2x + 7| \leq 5$.
20. $|3x - 4| > 10$.
21. $|4x - 11| < 25$.
22. $|2x - 6| \geq 24 - x$.
23. $|3x - 8| \leq 13 - x$.
24. $x - \dfrac{4}{x} > 3$.
25. $x - \dfrac{15}{x} > 2$.
26. $\dfrac{x}{5} < \dfrac{2x + 7}{x + 12}$.
27. $\dfrac{x}{4} > \dfrac{2x - 1}{3x + 1}$.
28. $\dfrac{(x^2 - 1)(x^2 - 9)}{(x^2 - 4)(x^2 - 16)} > 0$.
29. $\dfrac{(x^2 - 1)(x^2 - 9)}{(x^2 - 4)(x^2 - 16)} > 1$.

In Problems 30 through 33 find the domain of the given function in accordance with our standard agreement.

30. $\sqrt{x^2 + 10x + 21}$.
31. $\sqrt{\dfrac{1}{x^2 + 3x - 4}}$.
32. $\sqrt{(x^2 - 1)(x + 7)}$.
33. $\sqrt{4x - x^3}$.

Chapter 10

Parabolas, Ellipses, and Hyperbolas

1. The Conic Sections

As the name implies, a curve is called a *conic section* if it can be obtained as the intersection curve of a plane and a cone (see Figure 1). A *cone* is a 3-dimensional surface generated in the following way. Given a plane curve C, and a point P, not in the plane of the curve, the cone generated by these two items (the point P and the curve C) is the collection of all the points that lie on some line through P and some point of C.

For the most general cone, the curve C may be anything we like, but in this book, the curve C will always be a circle and the point P will always be on a line through the center of the circle and perpendicular to the plane of the circle. In this case the cone is called *a right circular cone*, but we will use the word *cone* to mean a right circular cone. The point P is called the *vertex* of the cone, and the curve C is called a *director* of the cone.

The mathematical cone is the idealization of the usual ice cream cone, but there are two important differences (a) the cone extends to infinity and (b) it extends in both directions from the vertex.

Figure 1 shows three different possibilities for the position of the plane that intersects the cone. In (a) the intersection curve of the plane and the cone is called an *ellipse*. In (b) this

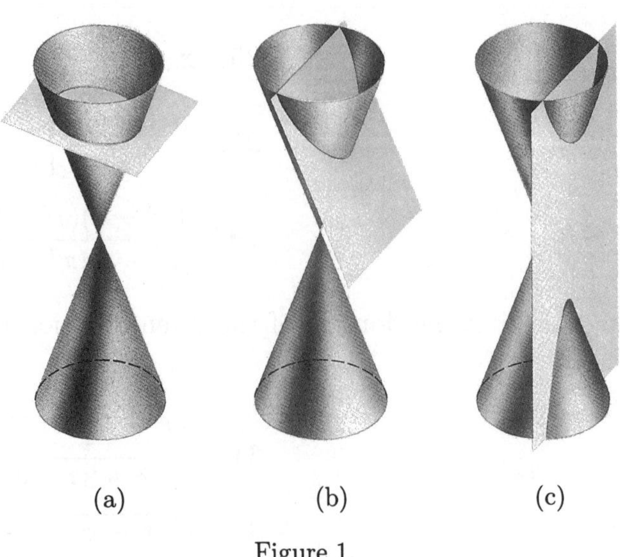

(a) (b) (c)

Figure 1.

intersection curve is a *parabola*, and in (c) the intersection curve is a *hyperbola*. In Exercise 1 we will look at a few other possibilities for the intersection set.

The ancient Greek mathematicians discovered these three conic sections, and made remarkable progress in finding their most interesting properties. In fact their work on the conic sections occupies a prominent place among the beautiful achievements of that period (roughly 600 B.C. to 500 A.D.)

During this period purely geometric methods were used to study the conic sections. However, when analytic geometry was introduced (in the seventeenth century) it became apparent that this subject offered a simpler method for investigating the conic sections, and this is the method that we will use in the rest of these booklets. In the next three sections we will use the rectangular coordinate system (Analytic Geometry) to define the parabola, the ellipse, and the hyperbola, and obtain their elementary properties.

A hundred years ago, the student who majored in mathematics studied the conic sections for an entire semester. But mathematics has grown so much and so many new subjects now demand our attention that the serious study of the conic sections has almost disappeared, and today the subject is treated very lightly (perhaps in one week).

Exercise 1

1. Can a circle be a conic section?

2. Can the set consisting of just one point be a conic section?

3. Can a pair of intersecting lines be a conic section?

4. Can a pair of parallel lines be a conic section?

5. Can a single line be a conic section?

---

2. The Parabola

The correct place to start is

Definition 1. The set of all points P that are equidistant from a fixed point F (called the *focus*) and a fixed line D (called the *directrix*) is called a *parabola*.

To find a nice easy equation for the parabola we will place the focus and the directrix in very convenient positions. But suppose that we are not allowed to move the focus and the directrix. In this case we will move the coordinate axes to a suitable position. In either case (see Figure 2) we locate the focus and the directrix so that the y-axis passes through

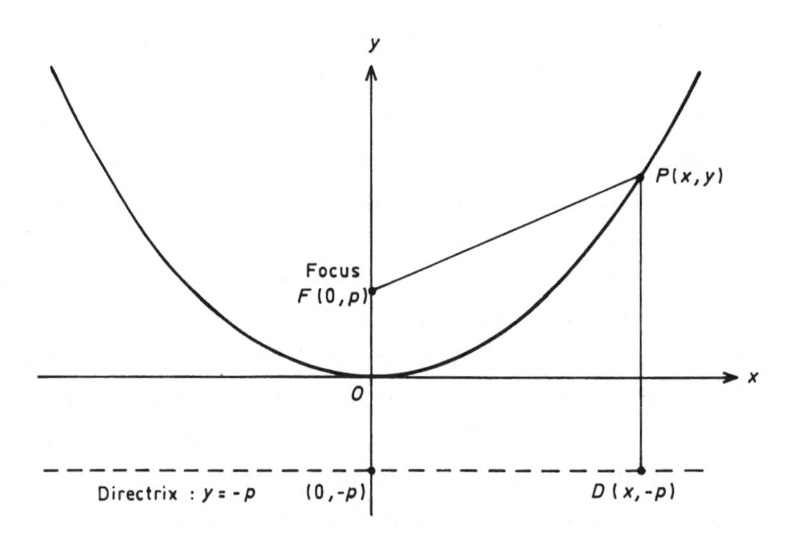

Figure 2.

the focus and is perpendicular to the directrix. Further, we can select the origin so that it bisects the line segment from the focus to the directrix with the focus above the x-axis (and consequently the directrix is below the x-axis and parallel to it). As a result of these choices of position, the focus will have the coordinates $(0, p)$ where $p > 0$ is some positive number. The directrix will meet the y-axis at the point $(0, -p)$ since the origin bisects the line segment from the focus to the directrix. Further, the directrix is perpendicular to the y-axis and so the equation of the directrix is $y = -p$.

Now let $P(x, y)$ be any point on the parabola. The distance from the point P to the focus F is given by

$$|PF| = \sqrt{(x-0)^2 + (y-p)^2}\,. \tag{1}$$

The distance from the point P to the directrix (measured along a line perpendicular to the directrix) is given by

$$|PD| = \sqrt{(x-x)^2 + (y-(-p))^2}\,. \tag{2}$$

The point P is on the parabola if and only if the two distances are equal. The same is true if the squares are equal. Using the squares, Equations (1) and (2) yield

$$(x-0)^2 + (y-p)^2 = (x-x)^2 + (y-(-p))^2\,,$$

or

$$x^2 + y^2 - 2py + p^2 = 0 + y^2 + 2py + p^2\,,$$

or

$$\boxed{x^2 = 4py\,,} \qquad p > 0\,. \tag{3}$$

Theorem 1. *The graph of Equation* (3) *is a parabola with the focus at* $(0, p)$. *The directrix is the line* $y = -p$ *and the distance from the focus to the directrix is* $2p$.

The intersection point of the parabola and the line through the focus perpendicular to the directrix is called the *vertex* of the parabola. We have proved that the vertex of the parabola (3) is at the origin, and that p is the distance from the focus to the vertex. When the parabola is located so that the focus is on the y-axis and the directrix is parallel to the x-axis and below it, and both have the same distance from the origin (see Figure 2), we say that the parabola is in *standard position*. We have proved that if the parabola is in standard position it will have an equation of the form (3) with $p > 0$.

Example 1. Give an equation for the parabola with focus at $(0, 2)$ and directrix $y = -2$.

Solution. For this parabola $p = 2$, so Equation (3) becomes

$$x^2 = 8y \quad \text{or} \quad y = \frac{x^2}{8} . \qquad ▶▶$$

Example 2. Find the distance from the focus to the vertex for the parabola $y = 1.5x^2$, and sketch a graph of this equation.

Solution. To compare the equation $y = 1,5x^2$ with Equation (3) we divide both sides by 1.5 to obtain $x^2 = y/1.5 = y/(3/2) = 2y/3$. When we compare this equation with Equation (3) we see that we must have $4p = 2/3$ or $p = 2/12 = 1/6$. Thus the focus is at $(0, 1/6)$, the directrix is the line $y = -1/6$, and the distance from the focus to the vertex at $(0, 0)$ is $1/6$. The student should make a sketch of $y = 1.5x^2$ and locate the focus and the directrix. ▶▶

Musicians would regard Equation (3) as the theme of this section. Now for some variations on this theme.

We say that the graph of Equation (3) is a parabola that opens upward. If we reflect this curve in the x-axis we obtain a parabola that opens downward and the equation is

$$x^2 = -4py , \quad p > 0 . \tag{4}$$

If we rotate the graph of Figure 2 about the origin in a clockwise direction 90° we obtain a parabola that opens to the right. The equation for such a parabola is

$$y^2 = 4px , \quad p > 0 . \tag{5}$$

If the rotation is 90° counterclockwise we obtain a parabola that opens to the left. The equation for such a parabola is

$$y^2 = -4px , \quad p > 0 . \tag{6}$$

Finally, we would like an equation for a parabola when the vertex is at some point (h, k) not the origin. We can obtain this equation by moving the curve from its standard position to a new place so that its vertex is at the origin in a *new coordinate system*. We use cap letters (X, Y) for the new coordinate system. The transformation from the old (x, y) coordinate system to the new (X, Y) system is pictured in Figure 3. Thus each point has

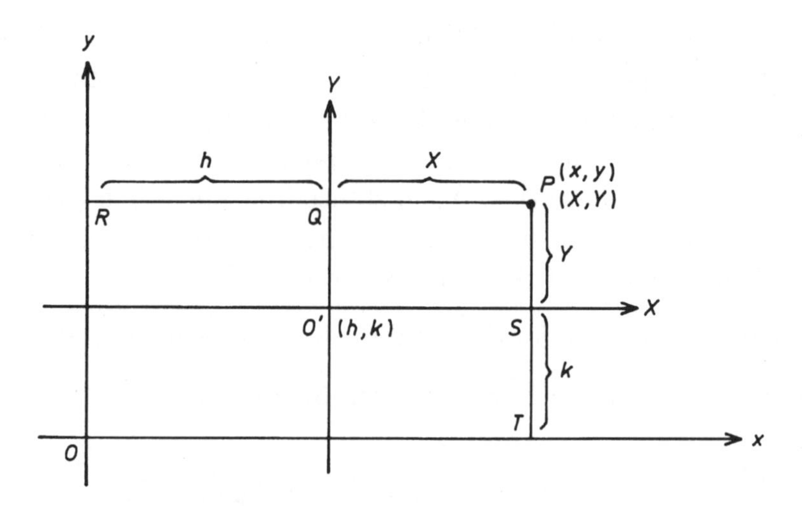

Figure 3.

TWO SETS of coordinates (x, y) and (X, Y) and it is clear from the picture that these coordinates are related by the set of equations

$$x = X + h, \quad y = Y + k, \tag{7}$$

or after subtracting h and k

$$X = x - h, \quad Y = y - k. \tag{8}$$

These equations, (7) and (8) are known as the equations for the translation of axes. They were obtained from a picture in which all of the letters represent positive numbers. However if we use directed distances and recall that the coordinates of a point are DIRECTED DISTANCES, then it will be clear that Equations (7) and (8) are correct for ANY PAIR of numbers (h, k). Thus we have proved

Theorem 2. *When the coordinate axes are translated so that the origin in the new system (X, Y) has coordinates (h, k) in the original (x, y) system, then the coordinates of a fixed point are related by Equations* (7) *and* (8).

We can also obtain equations for rotating the axes about the origin, but these equations are more complicated. They will not be used in these booklets, and so we omit them.

Equation set (8) is easy to memorize. We merely ask where is the origin $(0, 0)$ in the (X, Y) system. Answer: use $x = h$ and $y = k$ in equation set (8) since the origin in the (X, Y) system is at $(0, 0)$. The (X, Y) origin is at (h, k) in the (x, y) system.

Example 3. Find an equation for a parabola that opens upward, has its vertex at $(3, 2)$ and has its focus 5 units above the vertex.

Solution. Here $p = 5$, so the equation in the (X, Y) system is

$$X^2 = 4pY = 4(5)Y = 20Y. \tag{9}$$

Then equation set (8) gives $X = x - 3$ and $Y = y - 2$. Thus, (9) becomes

$$(x-3)^2 = 20(y-2)\,. \tag{10}$$

If we want to express y as a function of x, we solve Equation (10) for y and obtain

$$y = \frac{1}{20}(x-3)^2 + 2\,.$$ ▶▶

Example 4. Find an equation for the parabola that opens to the left, has the vertex at $(4,-1)$ and has the line $x = 7$ as its directrix. Where is the focus of this parabola?

Solution. Since the distance from the directrix to the vertex is $7 - 4 = 3$ we have $p = 3$. Thus in the (X,Y) system the equation is $Y^2 = -4pX = -4(3)X = -12X$. But from equation set (8)

$$X = x - h = x - 4 \quad \text{and} \quad Y = y - k = y - (-1) = y + 1\,.$$

Thus, $Y^2 = -12X$ gives $(y+1)^2 = -12(x-4)$. The focus is p units to the left of the vertex so the x coordinate is $4-3 = 1$. The vertex is at $(1,-1)$. The energetic student should sketch this parabola from the equation $(y+1)^2 = -12(x-4)$. ▶▶

Example 5. Prove that the graph of $y = x^2 + 3x - 1$ is a parabola. Find its vertex and focus.

Solution. Following the method of solving the general quadratic equation, we complete the square by adding (to both sides) the square of one-half the second coefficient. This gives

$$y + (3/2)^2 = x^2 + 3x + (3/2)^2 - 1\,,$$

or

$$y + (3/2)^2 = (x + 3/2)^2 - 1\,,$$

or

$$y + (3/2)^2 + 1 = y + 13/4 = (x + 3/2)^2\,. \tag{11}$$

If we use $X = x + 3/2$, and $Y = y + 13/4$, Equation (11) becomes $Y = X^2$ so in the (X,Y) system $4p = 1$, and $p = 1/4$, the vertex is at $(0,0)$ and the focus is at $(0,1/4)$. If we return to the original system, the (x,y) system, the graph is a parabola with vertex at $(-3/2,-13/4)$ and focus at $(-3/2,-3)$. ▶▶

The method of completing the square in Example 5 will give

Theorem 3. *If A is not zero, the graph of*

$$\boxed{y = Ax^2 + Bx + C\,,} \tag{12}$$

is a parabola.

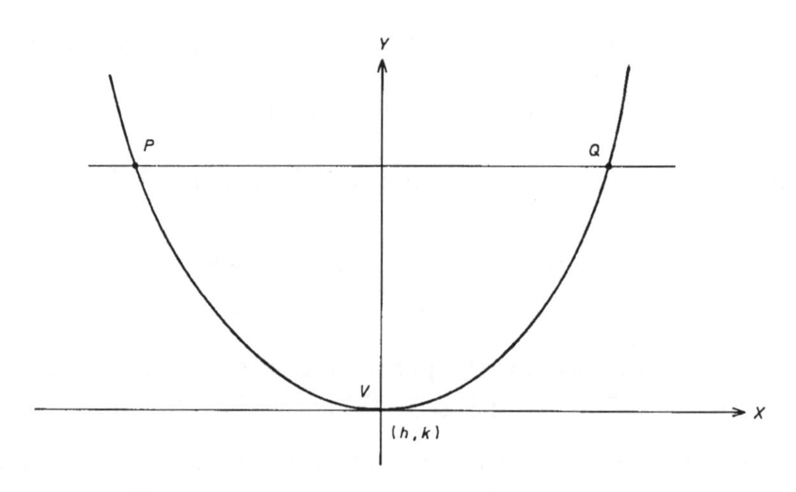

Figure 4.

A curve is said to be *symmetric* with respect to an axis if the portion of the curve on one side of the axis is the mirror image of the portion on the other side of the axis.

For a parabola, the line through the focus and perpendicular to the directrix is called the *axis of the parabola.*

A parabola is always symmetric with respect to its axis. This means that if a line perpendicular to the axis meets the parabola in two distinct points P and Q, then the axis bisects the line segment PQ, see Figure 4.

Exercise 2

(A)

In Problems 1 through 4, find the focus and the directrix of the given parabola.

1. $x^2 = 4y$. **2.** $y = x^2$. **3.** $y = 4x^2$. **4.** $y = x^2/32$.

5. Suppose that the focus is on the x-axis at $(p, 0)$, with $p > 0$ and the directrix is the vertical line $x = -p$. Prove that $y^2 = 4px$ is an equation for this parabola.

6. Suppose that the focus of the parabola is on the y-axis but below it so that it has coordinates $(0, -p)$ with $p > 0$. Suppose further that the directrix is the line $y = p$. Prove that in this situation $x^2 = -4py$ is an equation for the parabola.

In Problems 7 through 12, use the results of Problems 5 and 6 to find the focus and the directrix of the given parabola.

7. $y^2 = 4x$. **8.** $x^2 = -4y$. **9.** $y^2 = -4x$.

10. $y = -x^2/8$. **11.** $x = -y^2$. **12.** $7y = -5x^2$.

13. Graph (by hand) several of the parabolas given in Problems 1 through 4, and 7 through 12.

14. Find an equation for the parabola with focus at $(4, 5)$ and directrix $y = 1$.

15. Do Problem 14 if the focus is $(-2, -7)$ and the directrix is the y-axis.

***16.** Prove that for the parabola given in Theorem 3 the focus, vertex, and directrix are given by: Focus: $(-B/2A, (4AC - B^2)/4A + p)$, Vertex: $(-B/2A, (4AC - B^2)/4A)$, Directrix is the line: $y = (4AC - B^2)/4A - p$, where $p = 1/4A$.

17. Use the results in Problem 16 to find p, the vertex, the focus and the directrix for the parabola $8y = x^2 - 8x + 40$.

18. Do Problem 17 for the graph of $2y = 3x^2 + 4x - 5$.

19. Do Problem 17 for the graph of $y = -x^2 - 4x + 2$.

20. Use the definition of a parabola to prove that a parabola is always symmetric with respect to its axis, see Figure 4.

(B)

In Problems 1 through 8 put the graphs of the six different equations, all on the same coordinate axes on your monitor

1. Graph $y = x^2/2 + C$ for $C = -4$, -2, 0, 2, 4, and 6.

2. Graph $y = Ax^2$ for $A = -1$, $-1/2$, 1/4, 1/2, 1 and 2.

3. Graph $y = x^2 - 2ax + a^2 + a$ for $a = -1$, 0, 1, 2, 3, and 4.

4. Graph $y = \sqrt{x - a}$ for $a = -2$, 0, 2, 4, 6, and 8.

5. Graph $y = x^2 - 4bx + b(1 + 4b)$ for $b = -2$, -1, 0, 1, 2, and 3.

6. Graph $y = xa$ for $a = -1$, $-1/3$, 0, 1/3, 2/3, and 1.

7. Graph $y = xa$ for $a = 1$, 4/3, 5/3, 2, 7/3, and 8/3.

8. Graph the CUBIC polynomial $y = x(x^2 - a^2)/2$ for $a = 0$, 1.0, 1.5, 2, 2.5, and 3.0.

---

3. The Ellipse

As in Section 2 we start with

Definition 2. The set of all points P such that the sum of its distances $|PF_1| + |PF_2|$ from two fixed points F_1 and F_2 is a constant is called an *ellipse*. Each of the two points F_1 and F_2 is called a *focus* of the ellipse, see Figure 5.

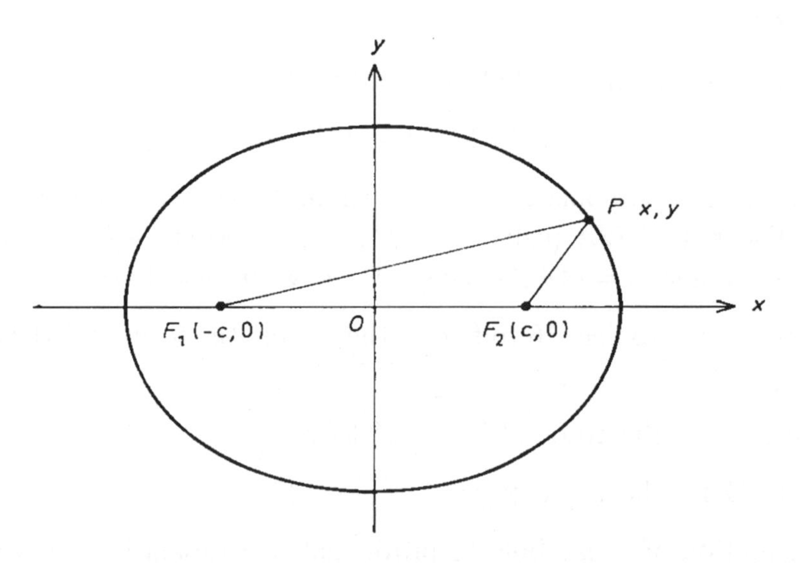

Figure 5.

As before, the task of finding an equation for an ellipse is simplified if we select the coordinate system judiciously. This selection is also illustrated in Figure 5. We let the x-axis pass through the two foci F_1 and F_2, and we let the y-axis bisect the line segment F_1F_2. As indicated in Figure 5 the coordinates of the foci are $(-c, 0)$ and $(c, 0)$, where $c > 0$ is the distance from the focus F_1 to the origin. When the ellipse is in this position (see Figure 5) we say that the ellipse is in *standard position.*

It is convenient to select $2a$ for the constant sum of the distances of $P(x, y)$ from the two foci. Then from the definition of an ellipse the point $P(x, y)$ is on the ellipse if and only if

$$|PF_1| + |PF_2| = 2a\,. \tag{13}$$

Using the formula for distances, Equation (13) yields

$$\sqrt{(x+c)^2 + y^2} + \sqrt{(x-c)^2 + y^2} = 2a\,. \tag{14}$$

To simplify this equation we must remove the radicals. One way to do this is to transpose the first radical to the right side of the equation, put $-2a$ on the left side, and square both sides thus

$$(\sqrt{(x-c)^2 + y^2} - 2a)^2 = (-\sqrt{(x+c)^2 + y^2}\,)^2\,,$$

$$x^2 - 2cx + c^2 + y^2 - 4a\sqrt{(x-c)^2 + y^2} + 4a^2 = x^2 + 2cx + c^2 + y^2\,. \tag{15}$$

Now x^2, c^2, and y^2 cancel on both sides. We move $-2cx$ to the right side and divide by 4. Then we have

$$a^2 - a\sqrt{(x-c)^2 + y^2} = cx\,. \tag{16}$$

Now subtract a^2 from both sides and square again. This will give

$$(-1)^2 a^2 (x^2 - 2cx + c^2 + y^2) = (cx - a^2)^2\,, \tag{17}$$

$$a^2x^2 - 2a^2cx + a^2c^2 + a^2y^2 = c^2x^2 - 2a^2cx + a^4\,. \tag{18}$$

We cancel the common terms $-2a^2cx$, move c^2x^2 to the left side, and move a^2c^2 to the right side. This gives

$$a^2x^2 - c^2x^2 + a^2y^2 = a^4 - a^2c^2\,,$$

or

$$(a^2 - c^2)x^2 + a^2y^2 = a^2(a^2 - c^2)\,. \tag{19}$$

Finally, if $a \neq c$ so we are not dividing by 0, we divide both sides by the right side of (19) and find that

$$\frac{x^2}{a^2} + \frac{y^2}{a^2 - c^2} = 1\,. \tag{20}$$

If we know that $a > c > 0$, we can introduce $b > 0$, by setting

$$b^2 = a^2 - c^2 \tag{21}$$

so that (21) simplifies to

$$\boxed{\frac{x^2}{a^2} + \frac{y^2}{b^2} = 1\,.} \qquad 0 < b < a\,. \tag{22}$$

The computation has been rather long but the simple form of Equation (22) makes all the labor well worthwhile. Equation (22) is called the *standard form for the equation of the ellipse.*

We must pick up an important loose end. To prove that b is well-defined, we must show that $a > c > 0$. Let P be any point on the ellipse. In the triangle inequality

$$|PF_1| + |PF_2| \geq |F_1F_2|\,, \tag{23}$$

the left side is $2a$ by definition, and the right side is $2c$ (see Figure 5). Therefore $2a \geq 2c > 0$, with equality if and only if the point P is on the line segment F_1F_2. Thus if $a = c$, the ellipse is the line segment F_1F_2. Such an ellipse is said to be a *degenerate ellipse.* In all other cases we have $a > c > 0$. We have proved

Theorem 4. *If the ellipse is in standard position it has an equation of the form* (22). *Conversely, the graph of any equation of the form* (22) *is an ellipse in standard position* (*see Figure* 5). *The foci are located at* $(-c, 0)$ *and* $(c, 0)$ *where*

$$\boxed{c = \sqrt{a^2 - b^2}\,.} \tag{24}$$

Strictly speaking, we have only proved the first half of the theorem, namely given the ellipse it has the Equation (22). To complete the proof we must prove that given the Equation (22) the graph is the ellipse of Figure 5. To do this last part we must show that all the steps from Equation (13) to Equation (22) can be reversed. This can be done, but there are some tricky points involved. For simplicity, we omit this part of the proof.

Example 1. Sketch the graph of $x^2/169 + y^2/25 = 1$. Where are the intercepts on the x- and y-axes? Where are the foci?

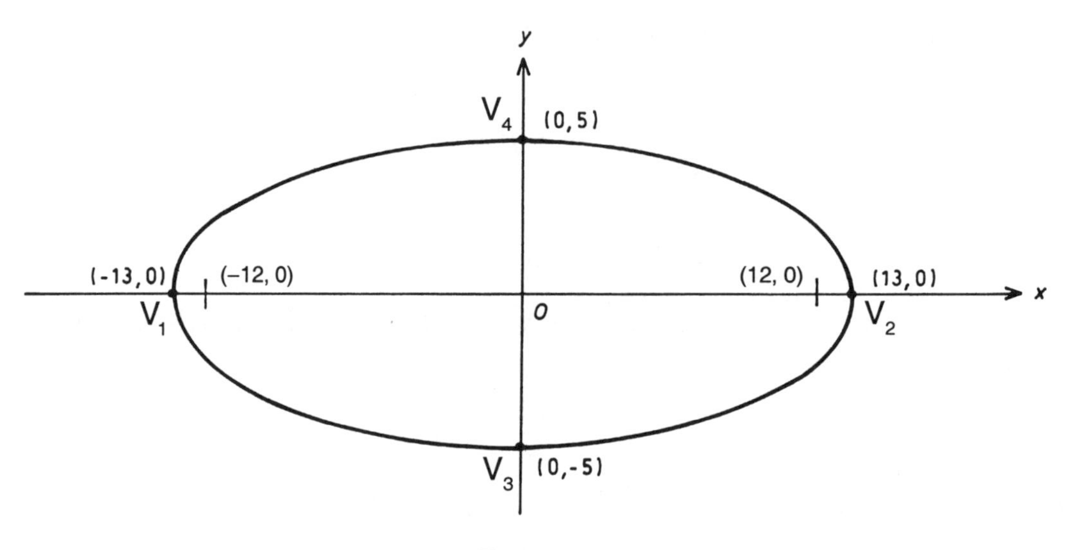

Figure 6.

Solution. A computer will give the graph very quickly. By hand, it is necessary to compute the coordinates of a few points. If we set $y = 0$ the equation gives $x^2 = 169$ or $x = 13$ and $x = -13$. Similarly, if $x = 0$, then from the equation $y^2 = 25$ or $y = 5$ and $y = -5$. These points: $(13, 0)$, $(-13, 0)$, $(0, 5)$, and $(0, -5)$ are easy to locate and are shown in Figure 6. One or two more points will convince the student that the curve has the shape shown in that figure. Equation (23) gives $c = \sqrt{13^2 - 5^2} = \sqrt{169 - 25} = \sqrt{144} = 12$. Therefore the foci are at $(-12, 0)$ and $(12, 0)$. ▶▶

Following the method of Example 1 we see that the points $V_1(-a, 0)$, $V_2(a, 0)$, $V_3(0, -b)$, and $V_4(0, b)$ are always on the graph of Equation (22). The first two are called the *vertices* of the ellipse. The segment V_1V_2 is called the *major axis* of the ellipse, and the segment V_3V_4 is called the *minor axis* of the ellipse. It is intuitively clear that the ellipse is symmetric with respect to the major axis and with respect to the minor axis.

The point of intersection of the major and minor axes is called the *center* of the ellipse. When the ellipse is in standard position the center of the ellipse is at the origin.

Exercise 3

(A)

In Problems 1 through 10 assume that the ellipse is in the standard position and that $a > b > 0$ in Equation (22).

1. What is the length of the major axis?

2. What is the length of the minor axis?

3. Prove that the ellipse is symmetric with respect to the major axis.

4. Do Problem 3 for the minor axis.

In Problems 5 through 10 find the equation of the ellipse in standard form that satisfies the given conditions.

5. The distance sum is 10 and the foci are $(\pm 3, 0)$.

6. The distance sum is 20 and the foci are $(\pm 6, 0)$.

7. The distance sum is 10 and the foci are $(\pm 2\sqrt{6}, 0)$.

8. The distance sum is 10 and the foci are $(\pm 1, 0)$.

9. The minor axis is 12, and the foci are $(\pm 5, 0)$.

10. The major axis is 18, and the foci are $(\pm 8, 0)$.

***11.** Is there an ellipse in which a, b, and c are proportional to 3, 2, and 1?

In Problems 12, through 14, find the foci for the ellipse.

12. $\dfrac{x^2}{26} + \dfrac{y^2}{12} = 1.$ **13.** $\dfrac{x^2}{4} + \dfrac{y^2}{3} = 1.$ ***14.** $\dfrac{x^2}{4} + 2y^2 = 1.$

***15.** Find an equation for the ellipse in standard position that passes through the points $(4, -1)$ and $(-2, -2)$.

***16.** Do Problem 15 for the points $(3, \sqrt{7},)$ and $(-\sqrt{3}, 3)$.

***17.** Find a general equation for an ellipse if the foci are at $(0, -c)$ and $(0, c)$.

18. Is a circle an ellipse?

***19.** Find a general equation for an ellipse if the center is at (h, k) and: (I) the major axis is parallel to the x-axis, (II) the major axis is parallel to the y-axis.

In Problems 20, 21, and 22 complete the square and put the equation in the form given in the answer to Problem 19. Then use this form to give: (a) the foci of the ellipse, (b) the vertices and (c) the axes of symmetry.

***20.** $9x^2 + 25y^2 - 90x - 150y + 225 = 0.$

***21.** $16x^2 + y^2 - 32x + 4y + 16 = 0.$

***22.** $25x^2 + 16y^2 + 100x - 192y + 276 = 0.$

23. Prove that the ellipse given by Equation (22) always lies inside the rectangle bounded by the lines $x = -a$, $x = a$, $y = -b$, and $y = b$. The only points on the boundary of the box are the intercepts on the x- and y-axes.

24. Prove that each ellipse of the family

$$\boxed{\frac{x^2}{a^2} + \frac{y^2}{a^2-1} = 1\,,} \qquad a > 1\,, \tag{25}$$

has the foci $(\pm 1, 0)$. The family is called a family of confocal ellipses.

25. What can you say about the family of ellipses

$$\boxed{\frac{x^2}{a^2-4} + \frac{y^2}{a^2} = 1\,,} \qquad a > 2? \tag{26}$$

(B)

In Problems 1 through 9, put the graphs of the six different equations on the same coordinate axes on your monitor. If the curve has a horizontal axis of symmetry, it is sufficient to graph only the upper half of the curve.

1. Graph $x^2/a^2 + y^2/4 = 1$, for $a = 1$, 2, 3, 4, 5, and 6.

2. Graph $x^2/4^2 + y^2/b^2 = 1$, for $b = 1$, 2, 3, 4, 5, and 6.

3. Graph $x^2/(4a^2) + y^2/a^2 = 1$, for $a = 1$, 2, 3, 4, 5, and 6.

4. Graph $(x-a)^2/4 + (y-a)^2 = 1$, for $a = -2$, -1, 0, 1, 2, and 3.

5. Graph $(x-2a)^2/9 + (y-a)^2 = 1$, for $a = -2$, -1, 0, 1, 2, and 3.

6. Graph $x^3 + y^3 = a^3$, for $a = 1$, 2, 3, 4, 5, and 6.

7. Graph $x^4 + y^4 = a^4$, for $a = 1$, 2, 3, 4, 5, and 6.

8. Graph $x^{1/2} + y^{1/2} = a$, for $a = 1, \sqrt{2}, \sqrt{3}$, 2, $\sqrt{5}$, and $\sqrt{6}$. Note, we must have x, $y \geq 0$.

9. Graph $x^a + y^a = 1$, for $a = 1/3$, 1/2, 1, 3/2, 5/2, and 3.

---

4. The Hyperbola

As in Section 3 we start with

Definition 3 (Hyperbola). The set of points P such that the difference of its distances from two fixed points F_1 and F_2 is a positive constant, is called a *hyperbola*. Each of the points F_1 and F_2 is called a *focus* of the hyperbola.

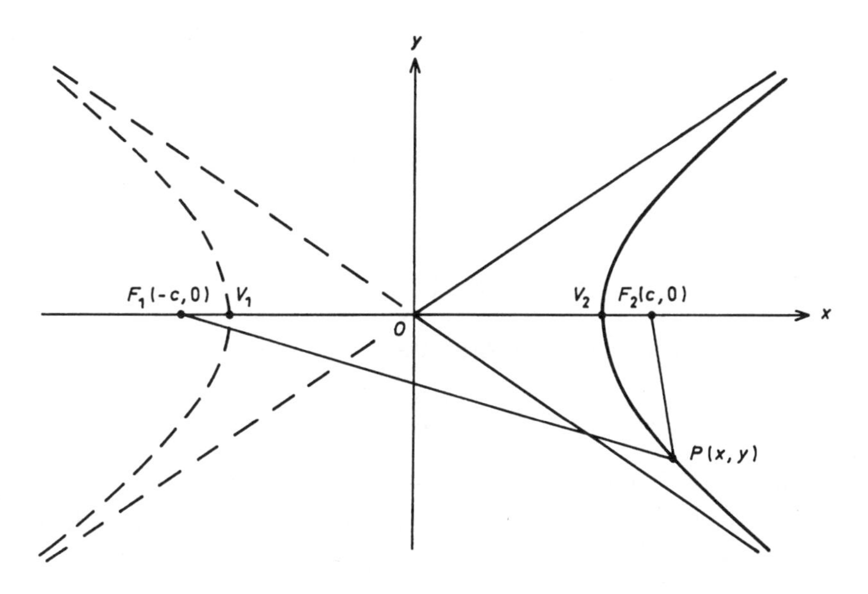

Figure 7.

Here we should observe that the order of the difference is not specified, so there are two possibilities for the subtraction. As in the case of the ellipse, we let the difference be $2a$. Then the point P is on the hyperbola if and only if P satisfies either

$$|PF_1| - |PF_2| = 2a\,, \tag{27}$$

or

$$|PF_2| - |PF_1| = 2a\,. \tag{28}$$

To find a simple form for the equation of a hyperbola we select our coordinate axes judiciously. As in the case of the ellipse we run the x-axis through the two foci, and we let y-axis bisect the line segment F_1F_2. Under these conditions we say that the hyperbola is in *standard position.* If the number $c > 0$ is suitably selected then the coordinates of the foci are $(-c, 0)$ and $(c, 0)$. These items are shown in Figure 7. This same figure also shows a hyperbola, but at this moment we are not certain that the graph is drawn correctly.

Let $P(x, y)$ be a point on the hyperbola, and assume that it satisfies Equation (27). Then by our distance formula we have

$$\sqrt{(x+c)^2+y^2} - \sqrt{(x-c)^2+y^2} = 2a\,. \tag{29}$$

We move the second radical in (29) to the right side and then square both sides. This gives

$$(x+c)^2 + y^2 = 4a^2 + 4a\sqrt{(x-c)^2+y^2} + (x-c)^2 + y^2\,,$$

or

$$x^2 + 2cx + c^2 + y^2 = 4a^2 + 4a\sqrt{(x-c)^2+y^2} + x^2 - 2cx + c^2 + y^2\,.$$

The common terms x^2, c^2, and y^2 on both sides cancel and when we combine the terms $2cx$, and divide both sides by 4, we have

$$xc = a^2 + a\sqrt{(x-c)^2 + y^2}\,,$$

or

$$xc - a^2 = a\sqrt{(x-c)^2 + y^2}\,. \tag{30}$$

Now square both sides and put the right side first. This gives

$$a^2x^2 - 2a^2cx + a^2c^2 + a^2y^2 = c^2x^2 - 2a^2cx + a^4\,. \tag{31}$$

Compare this equation with Equation (18) of the section on ellipses. This is exactly the same equation as Equation (18) that we obtained for the ellipse. All of the work from Equation (18) to Equation (20) is exactly the same, and thus we obtain Equation (20) which we renumber for convenience as

$$\frac{x^2}{a^2} + \frac{y^2}{a^2 - c^2} = 1\,. \tag{32}$$

Since Equations (20) and (32) are identical, what is the difference between the ellipse and the hyperbola? Of course the graphs in Figures 5 and 7 are different and the definitions are completely different. But in Equation (20), the difference lies in the relation between a and c. When we considered the ellipse, the triangle inequality dictated that

$$a > c > 0\,. \tag{33}$$

Now for the hyperbola, we have either

$$|PF_1| - |PF_2| = 2a\,, \tag{27}$$

or

$$|PF_2| - |PF_1| = 2a\,. \tag{28}$$

Equation (27), a small picture, and the triangle inequality will convince the student that $|PF_1| \leq |PF_2| + |F_1F_2|$, or $|PF_1| - |PF_2| \leq |F_1F_2| = 2c$, so that

$$\boxed{0 < a < c\,.} \tag{34}$$

The interested reader can prove that Equation (28) also gives (34). So here is the difference. For the ellipse $0 < c < a$ and for the hyperbola $0 < a < c$. Hence, in Equation (32) we can put $b^2 = c^2 - a^2$ because b is well-defined as a positive number. Then (32) becomes

$$\boxed{\frac{x^2}{a^2} - \frac{y^2}{b^2} = 1\,,} \tag{35}$$

since $a^2 - c^2 = -b^2$. We have almost proved

Theorem 5. *If the hyperbola is in standard position, then it has an equation of the form* (35)*. Conversely, the graph of any equation of the form* (35) *is a hyperbola in standard position. The foci are located at* $(-c, 0)$ *and* $(c, 0)$*, where*

$$c = \sqrt{a^2 + b^2}\,. \tag{36}$$

Strictly speaking, we have only proved a part of the theorem, namely given a point P that satisfies the condition (27), then the coordinates satisfy Equation (35). To complete the proof we must also consider those points that satisfy the condition (28). Further we must prove that if the coordinates of a point $P(x, y)$ satisfy Equation (35), then the point P satisfies either condition (27) or (28). To do this last part we must show that all the steps from Equation (29) to Equation (36) can be reversed. This can be done, but there are some tricky points involved. For simplicity, we omit all of these items, because the method is now clear. With sufficient sweat, the proof can be completed.

Example 1. Sketch the graph of $x^2/16 - y^2/9 = 1$. Where are intercepts on the x- and y-axes? Where are the foci?

Solution. A computer will give the graph very quickly. By hand, it is necessary to compute the coordinates of a few points. If we set $y = 0$ the equation gives $x^2 = 16$ or $x = 4$ and $x = -4$. So the intercepts on the x-axis are $(-4, 0)$ and $(4, 0)$. But if we set $x = 0$, then the equation gives $y^2 = -9$ and hence there is no real y. Thus the hyperbola has no y-intercepts (or the hyperbola does NOT meet the y-axis). One or two more computed points will convince the student that the curve has the shape shown in Figure 7. Equation (36) gives $c = \sqrt{4^2 + 3^2} = \sqrt{16 + 9} = \sqrt{25} = 5$. Therefore the foci are at $(-5, 0)$ and $(5, 0)$. ►►

For the moment let us throw away rigor (absolute proofs) and use our intuition (our feelings for mathematics). Suppose that x or y is very large. Then in Equation (35) the effect of the "1" on the right side is very small and can be neglected. If we drop the "1" we can put (35) in the form

$$\frac{y^2}{b^2} = \frac{x^2}{a^2} \quad \text{or} \quad y = \pm\frac{b}{a}\,x\,. \tag{37}$$

Equation (37) determines two straight lines (shown in Figure 7). The hyperbola gets closer and closer to these lines as the curve gets farther and farther from the origin. Such lines are called *asymptotes* and the hyperbola in standard position has the two asymptotes $y = \pm bx/a$. These are the lines in Figure 7.

The line through the foci of the hyperbola is called the *axis* of the hyperbola. The perpendicular bisector of the segment F_1F_2 is called the *conjugate axis* of the hyperbola. It is intuitively clear that the hyperbola is symmetric with respect to either of these axes. The intersection points of the hyperbola and the line F_1F_2 are the *vertices* of the hyperbola. These points are labeled V_1 and V_2 in Figure 7. The intersection of the axis and the conjugate axis is called the *center* of the hyperbola. As may be expected, when the hyperbola is in standard position the center is at the origin.

Exercise 4

(A)

In Problems 1 through 10 assume that the hyperbola is in the standard position.

1. What is the distance between the two vertices?

2. Prove that a hyperbola is symmetric with respect to its axis.

3. Do Problem 2 for the conjugate axis of the hyperbola.

In Problems 4 through 6 find the equation of the hyperbola in standard position that satisfies the given conditions.

4. The distance difference is 16 and the foci are at $(\pm 10, 0)$

5. The distance difference is 32 and the foci are $(\pm 20, 0)$.

6. The distance difference is 10 and the foci are $(\pm 6, 0)$.

In Problems 7 through 10 find the foci of the hyperbola.

7. $x^2/9 - y^2/16 = 1$.

8. $x^2/2 - y^2/7 = 1$.

9. $x^2 - 2y^2 = 6$.

10. $2x^2 - 3y^2 = 5$.

11. Find the asymptotes for the graph of the equations given in: (a) Prob. 7, (b) Prob. 8, (c) Prob. 9, and (d) Prob. 10.

12. Find an equation for the hyperbola in standard position that has foci at $(\pm 2, 0)$ and passes through the point $(2, 3)$.

13. Find a general equation for a hyperbola if the foci are on the y-axis at $(0, -c)$ and $(0, c)$

***14.** Find a general equation for a hyperbola if the center is at (h, k) and: (a) the line through the foci is parallel to the x-axis, and (b) the line through the foci is parallel to the y-axis.

***15.** Find a general formula for the asymptotes of the hyperbola given in Problem 14.

In Problems 16, 17, and 18 use the method of completing the square, together with the formulas given in the answer to Problem 14 to find: (a) the foci, (b) the vertices, and (c) the axes of symmetry for the given hyperbola.

***16.** $16y^2 - 9x^2 - 64y - 54x = 161$.

***17.** $y^2 + 20 = x^2 + 10y + 4x$.

***18.** $6x^2 + 84x + 69 = 15y^2 + 90y$.

**19 Prove that the graph of $xy = K^2/2$ is a hyperbola by finding the equation of the hyperbola with foci at $(-K, -K)$ and (K, K) and distance difference $2K$, where $K > 0$.

(B)

In Problems 1 through 8, put the graphs of the six different equations on the same coordinate axes on your monitor. If the curve has a horizontal axis of symmetry, it is sufficient to graph only the upper half of the curve. Notice that we have interchanged x and y, so that. the graph gives an unbroken curve above the x-axis (the graph of Figure 7 has been rotated 90° about the origin).

1. Graph $y^2/a^2 - x^2/4 = 1$, for $a = 1$, 2, 3, 4, 5, and 6.
2. Graph $y^2/4^2 - x^2/b^2 = 1$, for $b = 1$, 2, 3, 4, 5, and 6.
3. Graph $y^2/(4a^2) - x^2/a^2 = 1$, for $a = 1$, 2, 3, 4, 5, and 6.
4. Graph $(y - a)^2/4 - (x - a)^2 = 1$, for $a = -2$, -1, 0, 1, 2, and 3.
5. Graph $(y - 2a)^2/9 - (x - a)^2 = 1$, for $a = -2$, -1, 0, 1, 2, and 3.
6. Graph $y^3 - x^3 = a^3$, for $a = -2$, -1, 0, 1, 2, and 3.
7. Graph $y^4 - x^4 = a^4$, for $a = 1$, 2, 3, 4, 5, and 6.
8. Graph $y^{1/2} - x^{1/2} = a$, for $a = 1$, 2, 3, 4, 5, and 6. Note we must have $x, y \geq 0$.

---

5. Eccentricity

The perceptive student must have been puzzled by the diverse definitions of the parabola, the ellipse, and the hyperbola. The parabola involves a ratio of distances with only one focus, the ellipse involves a sum of distances with two foci, and the hyperbola involves a difference of distances. There should be a uniform definition that includes all three of the conic sections. There is, and the unifying element is the eccentricity which we now explain.

Definition 4 (Eccentricity). Let $\mathcal{D}$ be a fixed line and let F be a fixed point. For a fixed number $e > 0$, let $\mathcal{G}$ be the collection of all points $P(x, y)$ such that

$$\frac{|PF|}{|PD|} = e, \tag{38}$$

where $|PF|$ is the distance from P to the point F, and $|PD|$ is the distance from P to the line $\mathcal{D}$. Then e is called the *eccentricity* of the graph $\mathcal{G}$, the point F is called a *focus* of the graph $\mathcal{G}$, and $\mathcal{D}$ is called a *directrix* of the graph $\mathcal{G}$, (see Figure 8).

Of course $\mathcal{G}$ is a conic section. This is the content of

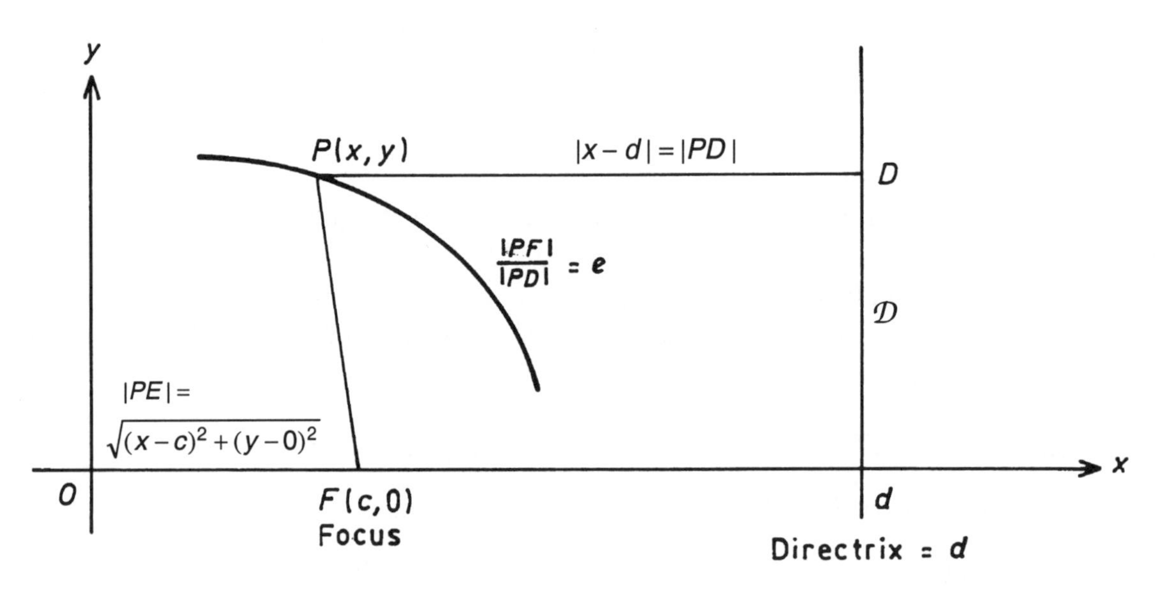

Figure 8.

Theorem 6. *In Definition 4, if $0 < e < 1$, then $\mathcal{G}$ is an ellipse, if $e = 1$, then $\mathcal{G}$ is a parabola, and if $1 < e$, then $\mathcal{G}$ is a hyperbola.*

Isn't this theorem a nice way of unifying the three conic sections?

Proof of Theorem 6. As usual we move the graph to a nice location (or if the graph cannot be moved we select the coordinate axes to suit our convenience). Whichever method is used, let the focus be on the x-axis at a point labeled $(c, 0)$ and let the directrix be perpendicular to the x-axis at $(d, 0)$ so that the equation of the directrix is $x = d$, (see Figure 8). Now, if $P(x, y)$ is any point in the plane, Equation (38) tells us that $P(x, y)$ is a point of the graph if and only if

$$|PF| = e|PD|\,. \tag{39}$$

Now, after squaring both sides of (39), and introducing the proper coordinates we have

$$(x-c)^2 + (y-0)^2 = e^2(x-d)^2\,,$$

or

$$x^2 - 2cx + c^2 + y^2 = e^2x^2 - 2e^2dx + e^2d^2\,,$$

or

$$(1-e^2)x^2 + y^2 + (2e^2d - 2c)x + (c^2 - e^2d^2) = 0\,. \tag{40}$$

The proof is essentially finished. The energetic reader may be waiting to complete the square in (40), but this labor is not needed to complete the proof of Theorem 6. We merely remark that Equation (40) is a quadratic equation in x and y without an xy term, so it must be the equation of a conic section. If $0 < e < 1$, then the coefficients of x^2 and y^2 are both

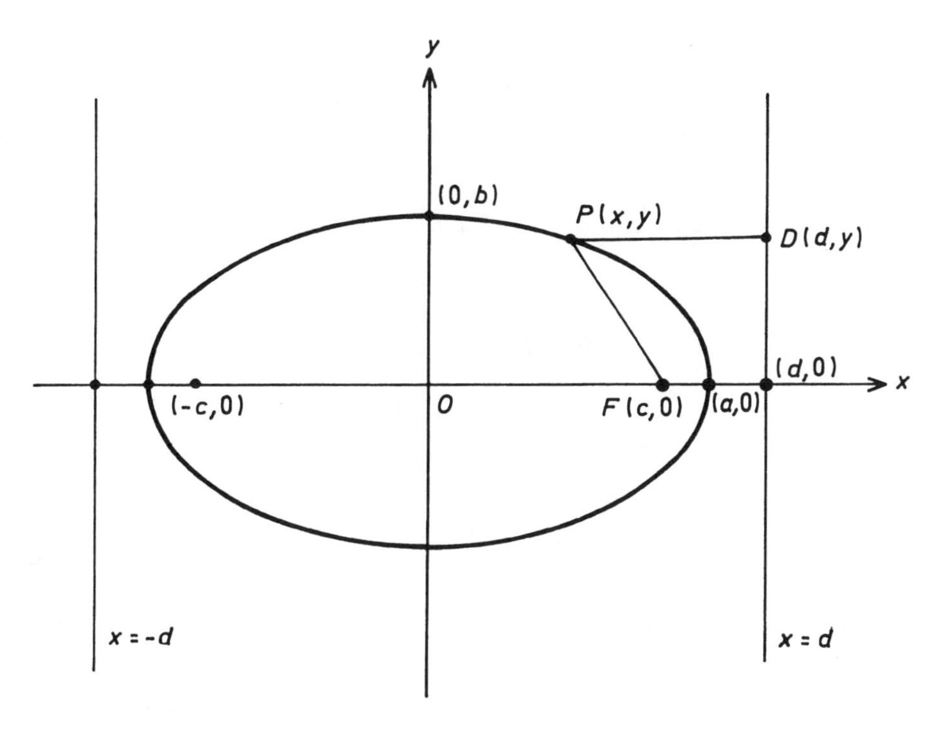

Figure 9.

positive and so this equation is the equation of an ellipse. If $e = 1$, then x^2 is missing and hence $\mathcal{G}$ is a parabola. Finally if $e^2 > 1$, then the coefficients of x^2 and y^2 have different signs and $\mathcal{G}$ is a hyperbola. ■

We could stop with the proof of Theorem 6, but the curious student will want to know the value of e for a given conic, and also the location of the directrix. For the parabola, this is easy because $e = 1$, and the location of the directrix is a part of the definition of a parabola. The other two conics require some labor.

Theorem 7. *If $a > b > 0$, set $c = \sqrt{a^2 - b^2}$. Then the ellipse*

$$\frac{x^2}{a^2} + \frac{y^2}{b^2} = 1 \tag{41}$$

has a focus at $F(c, 0)$ and a directrix at $x = d$, where

$$e = \frac{c}{a}, \quad \text{and} \quad d = \frac{a}{e}. \tag{42}$$

Further, e given in Equation (42), is the eccentricity.

Notice that $c < a$, so from (42) we have $0 < e < 1$ and $d > a$.

Proof of Theorem 7. Let $P(x, y)$ be any point on the ellipse (41). From Figure 9, we have

$$|PF|^2 = (x - c)^2 + (y - 0)^2 = x^2 - 2cx + c^2 + y^2\,.$$

If we use the value of y from Equation (41) and $c^2 = a^2 - b^2$, we will find that

$$|PF|^2 = x^2 - 2cx + c^2 + y^2 = x^2 - 2cx + c^2 + b^2\left(1 - \frac{x^2}{a^2}\right) \tag{43}$$

$$= x^2\frac{a^2 - b^2}{a^2} - 2cx + c^2 + b^2$$

$$= x^2\frac{c^2}{a^2} - 2\frac{c^2}{a^2}\left(\frac{a^2}{c}\right)x + c^2 + (a^2 - c^2)$$

$$= x^2\frac{c^2}{a^2} - 2\frac{c^2}{a^2}\left(\frac{a^2}{c}\right)x + \frac{c^2}{a^2}\frac{a^4}{c^2}$$

$$= \frac{c^2}{a^2}\left(x^2 - 2\frac{a^2}{c}x + \frac{a^4}{c^2}\right).$$

Now take the positive square root of the extremes in (43) to find

$$|PF| = \frac{c}{a}\left|x - \frac{a^2}{c}\right| = e\left|x - \frac{a}{e}\right| = e|x - d| = e|PD|. \tag{44}$$

■

Since the graph of Equation (41) is symmetric with respect to the y-axis, we see that the point $(-c, 0)$ is also a focus with the line $x = -d$ as the associated directrix (see Figure 9). Thus the ellipse has two foci, each with its own directrix.

What is the situation when $b > a > 0$? If we interchange the role of a and b we obtain the answer presented in

Theorem 8. *Suppose that in Equation* (41) *we have* $b > a > 0$. *Then the graph of* (41) *is an ellipse with its major axis on the y-axis. It has two foci at* $(0, -c)$ *and* $(0, c)$, *where* $c = \sqrt{b^2 - a^2}$. *Further, the eccentricity is* $e = c/b$ *and it has two directrices* $y = d$ *and* $y = -d$, *where* $d = b/e$.

For the hyperbola, in standard position, we have

Theorem 9. *Let* a, $b > 0$ *and set*

$$c = \sqrt{a^2 + b^2}, \quad e = \frac{c}{a}, \quad \text{and} \quad d = \frac{a}{e}. \tag{45}$$

Then the graph of

$$\frac{x^2}{a^2} - \frac{y^2}{b^2} = 1 \tag{46}$$

is a hyperbola with eccentricity e, *foci at* $(c, 0)$ *and* $(-c, 0)$ *and directrices* $x = \pm d$.

These items (except for e) are shown in Figure 10.

Proof of Theorem 9. By Definition 4 we must show that for any point $P(x, y)$ whose coordinates satisfy (46) we also have (38) or

$$|PF| = e|PD|. \tag{47}$$

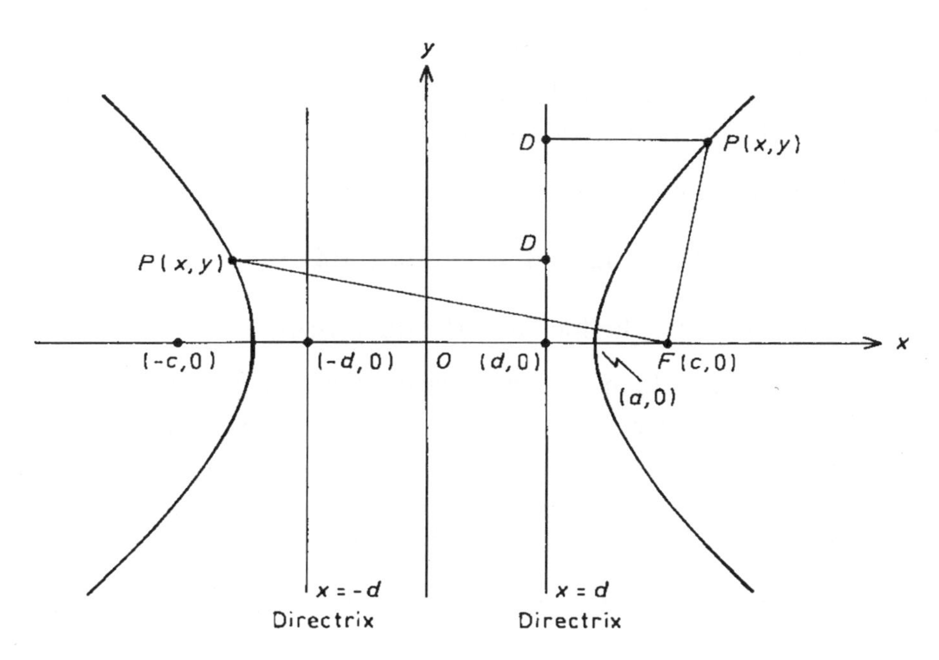

Figure 10.

Now, from (46) we have $y^2 = b^2(x^2/a^2 - 1)$. So our distance formulas give

$$|PF|^2 = (x-c)^2 + (y-0)^2 = x^2 - 2cx + c^2 + y^2 \tag{48}$$
$$= x^2 - 2cx + c^2 + b^2\left(\frac{x^2}{a^2} - 1\right).$$

Manipulations, similar to the ones for the ellipse in the proof of Theorem 7, will give

$$|PF|^2 = x^2\frac{a^2+b^2}{a^2} - 2cx + c^2 - b^2 = \frac{c^2}{a^2}\left(x^2 - 2\frac{a^2}{c}x + \frac{a^4}{c^2}\right).$$

Now take the positive square root of the extremes. This gives

$$|PF| = e|x - a^2/c| = e|x - a/e| = e|x - d| = e|PD|. \tag{49}$$

As in the case of the ellipse, the second focus and directrix follow by symmetry. ■

The student should note that for the ellipse $e < 1$, and hence $c = ea < a$. So the foci lie INSIDE the ellipse. But $d = a/e > a$ so the directrices lie OUTSIDE the ellipse (see Figure 9).

For the hyperbola, the opposite is true. We have $e > 1$, and $c = ea > a$. Further $d = a/e < a$. Thus the foci and directrices have the location (correctly) shown in Figure 10.

Example 1. Identify the graph of $4x^2 + y^2 = 1$. Find the eccentricity, the foci, and the directrices for this graph.

Solution. Since both coefficients, 4 and 1, are positive, the graph is an ellipse. To match the standard form, we must observe that $4 = 1/(1/4)$, so that $a^2 = 1/4$ and $b^2 = 1$. Since $b = 1 > 1/2 = a$ the foci are on the y-axis. From $c^2 = 1 - 1/4 = 3/4$, the foci are $(0, \pm\sqrt{3}/2)$. Now $e = c/a = (\sqrt{3}/2)/1 = \sqrt{3}/2$. Finally, we have $d = b/e = 1/(\sqrt{3}/2) = 2/\sqrt{3}$ and this is approximately $1.155 > 1$. The directrices are the horizontal lines $y = \pm 2/\sqrt{3}$. ▶▶

Example 2. Identify the graph of

$$25x^2 - 4y^2 - 150x - 16y + 109 = 0\,. \tag{50}$$

Find the center, the eccentricity, the foci, and the directrices.

Solution. Since the coefficients of x^2 and y^2 have different signs, we know that the graph is a hyperbola. To find the center etc., we must complete the square. The usual technique applied to (50) gives

$$\begin{aligned} 25(x^2 - 6x) \quad -4(y^2 + 4y) &= -109\,, \\ 25(x^2 - 6x + 9) - 4(y^2 + 4y + 4) &= -109 + 225 - 16 = 100\,, \\ 25(x - 3)^2 \quad -4(y + 2)^2 &= 100\,, \end{aligned}$$

$$\frac{(x-3)^2}{4} - \frac{(y+2)^2}{25} = 1\,. \tag{51}$$

(Weren't we lucky that the numbers turned out to be so nice?) From (51) the center is at $(3, -2)$, $c = \sqrt{25 + 4} = \sqrt{29}$ so the foci are at $(3 \pm \sqrt{29}, -2)$. Further the eccentricity is $e = c/a = \sqrt{29}/2$ and $d = a/e = 2/(\sqrt{29}/2) = 4/\sqrt{29}$ and the directrices are the two lines $x = 3 \pm 4/\sqrt{29}$ (see Figure 11). To find equations for the asymptotes, replace 1 by

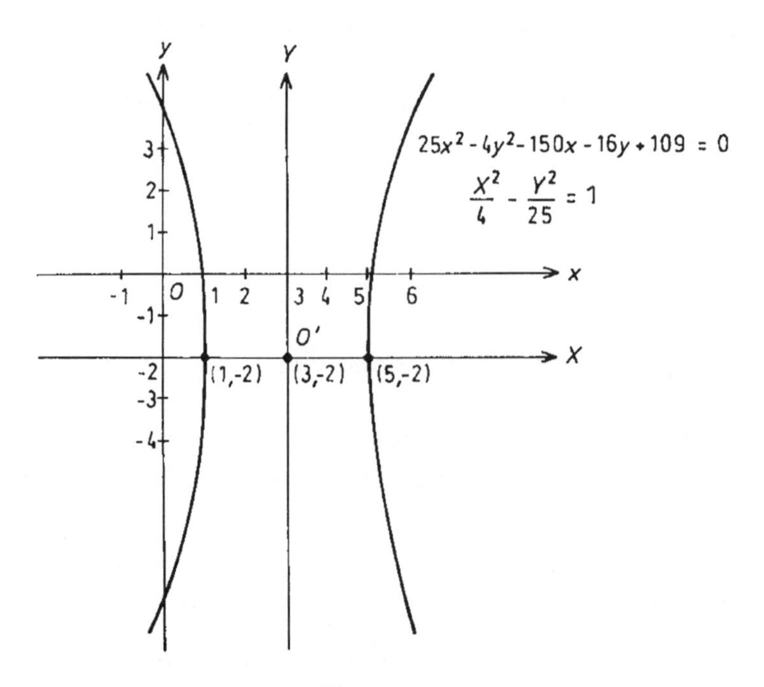

Figure 11.

0 in Equation (51) and solve for y. This operation should give $y+2 = \pm 5(x-3)/2$ or $y = -2 \pm 5(x-3)/2$.

Thus the equation of one asymptote is $y = 5x/2 - 19/2$. For the second asymptote we have $y+2 = -5(x-3)/2$ or

$$y = -\frac{5}{2}x + \frac{11}{2}.$$

Exercise 5

(A)

In Problems 1 through 4, find the focus, the vertex, and the directrix of the given parabola.

1. $y^2 - 4y + 12 = 8x$.

2. $x^2 + 2x + 16y + 33 = 0$.

3. $x + 20y^2 + 40y + 27 = 0$.

4. $2y + 180x = x^2 + 7950$.

In Problems 5 through 10 find the eccentricity, the foci, and the directrices of the given conic section. Hint: Complete the square as in Example 2.

5. $3x^2 + 4y^2 - 16y = 92$.

6. $25x^2 + 16y^2 + 200x + 400 = 160y$.

7. $8x^2 + 32x + 23 = y(y+2)$.

8. $9y^2 + 96x = 16x^2 + 72y + 144$.

9. $2(y^2 - 6y + 3) = x(x+4)$

10. $4x(x-2) + 3y(y+2) = 41$.

11. Let r be the ratio of the minor axis to the major axis in an ellipse. Prove that $r^2+e^2 = 1$. Prove that if $e = 0$, then the conic section is a circle.

In Problems 12 through 19 find the equation of the conic section that is in standard form (center at the origin) and satisfies the given conditions.

12. Focus at $(10, 0)$, and directrix $x = 8$.

13. Eccentricity 5, and directrix $y = 2$.

14. Focus at $(3, 0)$, and eccentricity 1.5.

15. Focus at $(4, 0)$, and directrix $x = -9$.

16. Ellipse passing through $(2, 1)$ and $(1, 3)$.

17. Focus at $(\sqrt{5}, 0)$, and asymptotes $2y = \pm x$.

18. Major axis 6, and focus at $(2, 0)$.

19. Eccentricity $1/9$, and focus at $(1, 0)$.

---

6. The Intersection of Pairs of Curves

A point is said to be an *intersection point* of two curves if it lies on both curves. This means that the coordinates of the point satisfy both of the equations of the curves and therefore all of the usual algebraic operations can be used on both equations SIMULTANEOUSLY. Conversely, any number pair that satisfies both equations will be the coordinates of a point of intersection of the two curves. This will be illustrated in our examples.

Example 1. Find the solution for the system of equations

$$y = -x - 4\,, \tag{52}$$

$$y = 6 + 2x - x^2\,. \tag{53}$$

Solution. Since Equation (52) is the equation of a line and Equation (53) is the equation of a parabola, we are looking for the points of intersection of a line and a parabola. Thus we may expect (a) two solutions, (b) one solution, or (c) no solutions.

We use the expression for y from (52) in (53) and find

$$-x - 4 = 6 + 2x - x^2\,,$$

or

$$x^2 - 3x - 10 = 0\,. \tag{54}$$

Now $x^2 - 3x - 10 = (x - 5)(x + 2)$ so either $x = 5$ or $x = -2$. When we use these values in Equation (52) we find that $(-2, -2)$ and $(5, -9)$ are the only possible points of intersection. The graphs of (52) and (53) together with the points of intersection are shown in Figure 12.

Example 2. Find the number of points of intersection of the parabola

$$y = 6 + 2x - x^2\,, \tag{53}$$

and the line

$$y = -x + k \tag{55}$$

as a function of k.

Solution. First notice that when $k = -4$ these are the two equations in Example 1. Following the method used in Example 1, we substitute y from Equation (55) in Equation (53) and have

$$-x + k = 6 + 2x - x^2\,,$$

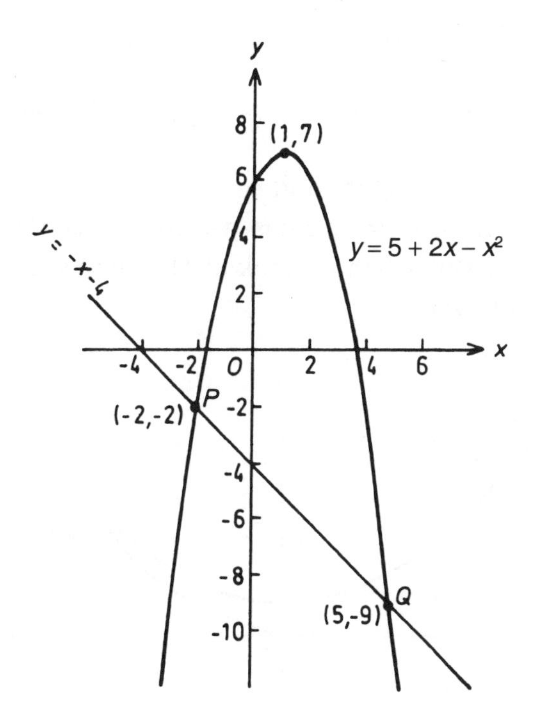

Figure 12.

or

$$x^2 - 3x + k - 6 = 0\,. \tag{56}$$

The quadratic formula gives

$$x = \frac{3 \pm \sqrt{9 - 4(k-6)}}{2} = \frac{3 \pm \sqrt{33 - 4k}}{2}\,. \tag{57}$$

When $k > 33/4$, the quantity under the radical is negative and so the roots of Equation (56) are not real numbers and there are no points of intersection. If $k < 33/4$, then Equation (56) has two roots and there are two points of intersection. Observe that in Figure 12, an increase in k moves the line upward. Thus when k is sufficiently large, the line will lie above the parabola and there will be no intersection points. The break point is $k = 33/4$ and when $k < 33/4$, the line meets the parabola in two points. It follows that when $k = 33/4$, the line just touches the parabola (is tangent to the parabola) and this occurs when Equation (56) has a double root. When $k = 33/4$, Equation (57) gives $x = 3/2$. Then Equation (53) gives $y = 27/4$ and the point $(3/2,\ 27/4)$ is the point at which $y = -x + 33/4$ is tangent to the parabola. ▶▶

Example 3. Find the solutions for the system

$$x^2 = 12(y-1)\,, \tag{58}$$

$$12y = x\sqrt{x^2 + 28}\,. \tag{59}$$

Solution. From Equation (58) we have $12y = x^2 + 12$. If we use this in (59) we have

$$x^2 + 12 = x\sqrt{x^2 + 28}\,. \tag{60}$$

To remove the radical we square both sides of (60) obtaining

$$x^4 + 24x^2 + 144 = x^2(x^2 + 28) = x^4 + 28x^2\,, \tag{61}$$

and hence $144 = 4x^2$. Therefore $x^2 = 36$, and so $x = 6$, or -6. From Equation (58) we find that when $x = 6$ or -6 we have $y = 4$. Thus we present the points $(6, 4)$ and $(-6, 4)$ as solutions of the given system. But when we look at the curves in Figure 13, we see only one point of intersection, and not two.

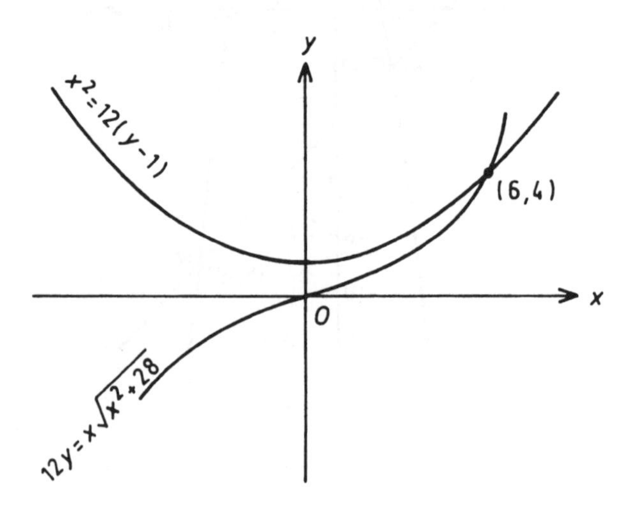

Figure 13.

How did this happen? We found two points of intersection, when Figure 13 shows clearly that there is only one such point. The quantity $\sqrt{x^2+28}$ is never negative by the definition of the radical sign. Thus the product $x\sqrt{x^2+28}$ is positive when x is positive, and is negative when x is negative. Hence, the graph of $y = x\sqrt{x^2+28}$ has the shape shown in Figure 13. Since the graph of $y = x^2/12 + 1$ lies above the x-axis, it is impossible for the two curves to meet at $(-6, 4)$. How did this error creep in? When we squared both sides of Equation (60) to obtain Equation (61) we introduced the possibility of extraneous roots. Recall that we can have $1 \neq -1$, but on squaring both sides we get the true equation: $1 = 1$. This is essentially what happened when we squared both sides of Equation (60) to get Equation (61). The ambitious student should substitute $x = -6$ in both equations. ▶▶

Exercise 6

(A)

In Problems 1 through 19 solve the given system of equations.

1. $y = 6x - 3 - x^2$,

$y = 2.$

2. $y = x^2 - 5x - 2$,

$y = 4.$

3. $y = 9 + 2x - x^2,$

$y = -3x + 13.$

4. $y = x^2 - 2x - 3,$

$y = 3 - x.$

5. $y = 1 + x^2,$

$y = 1 - 4x - x^2.$

6. $y = 2x^2 - 3x - 10,$

$y = x^2 - 2x + 10.$

7. $y = 2x - 4,$

$y = \sqrt{x^2 - 4x + 7}.$

8. $y = 2x,$

$y = \sqrt{x^2 + 3}.$

9. $y = 3x,$

$2y = \sqrt{9x^2 + 12}.$

***10.** $y = -x + 2,$

$y = x^3 - x^2 - 5x + 6.$

***11.** $y = x + 4,$

$y = x^3 - 3x^2 + x + 4.$

***12.** $y = 3x + 5,$

$y = x^3 - x^2 - 7x - 3.$

***13.** $2y = x + 3,$

$y = |x|.$

***14.** $2y = x + 5,$

$y = |x + 2|.$

***15.** $\dfrac{x^2}{18} + \dfrac{y^2}{8} = 1,$

$\dfrac{x^2}{3} - \dfrac{y^2}{2} = 1.$

16. $x^2 + y^2 = 16,$

$2x^2 - 3y^2 = 12.$

17. $x + y = 1,$

$x^2 + y^2 = 1.$

18. $x + y = 10,$

$x^2 + y^2 = 10.$

***19.** $y = -2x + 13,$

$y = 2x + 9/x^2.$

***20.** Find the largest value of r such that the line $x + y = r$ and the circle $x^2 + y^2 = r$ have a point in common.

Answers to Exercises in Volume 2

Chapter 6, Exercise 1

1. $-1, 1, 3, 5, 7, 23.$

2. $-8, -5, -2, 1, 4, 28.$

3. $9, 6, 5, 6, 9, 105.$

4. $10/9, 5/3, 2, 5/3, 10/9, 2/21.$

5. $-5/4, -2, -5, 10, 5/2, 5/14.$

6. $12, 6, 2, 0, 0, 72.$

7. $8, 3, 0, -1, 0, 3, 8.$

8. $12, 6, 2, 0, 0, 2, 6$

9. $-24, -6, 0, 0, 0, 6, 24.$

10. $0, 0, 0, 0, 24.$

11. $S = 6x^2.$

12. $V = x^3.$

13. $V = (S/6)^{3/2}.$

14. $A = \sqrt{3}\,x^2/4.$

15. $A = \sqrt{3}H^2/3.$

16. $2, 3, 6, 2, 8, 12.$

17. $4, 7, 12, 28, 20, 72.$

18. $3, 4, 9, 4, 4.$

19. Domain, all real numbers. Range, all real numbers.

20. Same as in Problem 19.

21. Domain, all real numbers. Range, all $y \geq 5$.

22. Domain, all real numbers. Range, all y for which $0 < y \leq 2$.

23. Domain, all real numbers except $x = 2/3$. Range, all y except $y = 0$.

24. Domain, all real numbers. Range, all $y \geq -1/4$.

25. Domain, all real numbers. Range, all $y \geq -1$.

26. Domain, all $x > 0$. Range, All $S > 0$.

27. Domain, all $x > 0$. Range, all $V > 0$.

28. Domain, all $x > 0$. But the reply "all real numbers" may be acceptable. Range, all integers $n \geq 0$.

29. $f(x) = x(x^2-1)(x^2-4)(x^2-9)(x^2-16)$, $f(5) = 362,880$.

30. $g(1) = -3$, $g(4) = 21$, $g(5) = 33$. No, $-3 + 21 \neq 33$.

31. $c^2 + 7c + 3$.

32. $(9+x)/(7-x)$.

33. $(9+z)/(5-z)$.

34. $(9+a)/(7-a)$.

35. $y^4 - 7y^2 + 10 = (y^2-2)(y^2-5)$.

36. $28x^2 - 48x - 65$.

37. $9x^2 - 60x - 117$.

38. $-96x - 111$.

Chapter 6, Exercise 2

1. $\{0, 3\}$.

2. $\{0, -6, 6\}$.

3. $\{0, 13/4\}$.

4. $\{1, 6\}$.

5. $\{-2, -1, 0\}$.

6. $\{-1\}$.

7. $\{7\}$.

8. $x = (a+b)/2$.

9. No solution.

10. $\{-19\}$.

11. $\{12\}$.

12. $\{6\}$.

13. $x = (a+b)/2$.

14. $\{-5, -3\}$.

Chapter 6, Exercise 3

1. -3.

2. $-5/3$.

3. $-1/6$.

4. $35/12$.

5. $-11/2$.

6. $1/2$.

7. 1.

8. $-21/19$.

9. $-28/9$.

10. $-12/11$.

11. $16/9$.

12. $-8/13$.

13. $5/4$.

14. $17/8$.

15. $5/2$.

16. $x = \dfrac{5ab - a^2 - b^2}{2a+b}$, $(a, b) = (1, 2), (3, 2), (2, 4)$ respectively for problems 13, 14, 15.

17. $y = \dfrac{2 - xz}{x+z}$.

18. $y = \dfrac{xz - 2x - 6z}{4 - x - z}$.

19. $y = \dfrac{-x^2}{z(1-x^2)}$.

20. $y = x^2/4p$.

21. $y = \pm b\sqrt{1 - \dfrac{x^2}{a^2}}$.

22. $y = \dfrac{a^2 + b^2 + c^2}{2(a+b+c)}$.

23. $y = \dfrac{x(a^2 + b^2 + c^2)}{2(a+b+c)}$.

24. $y = b \pm \sqrt{r^2 - (x-a)^2}$.

25. $y = b \pm \sqrt{r^2 - (x-a)^2 - (z-c^2)^3}$.

Chapter 6, Exercise 4

1. Use the symbols from Example 1,

$$y = x + 40\,, \quad d_1 + d_2 = 2550\,, \quad d_1 + d_2 = 8x + 6y = 8x + 6(x + 40)$$
$$8x + 6x + 240 = 2550\,, \quad 14x = 2310\,,$$
$$\text{so } x = 2310/14 = 165 \text{ miles/hour}, \quad y = 205 \text{ miles/hour}\,.$$

2. Use the symbols from Example 1,

$$y = x + 70\,, \quad d_1 + d_2 = 1000\,, \quad d_1 + d_2 = 3x + 2y = 3x + 2(x + 70)$$
$$5x = 1000 - 140\,, \quad \text{thus } x = 860/5\,,$$
$$x = 172 \text{ mules/hour}, \quad y = 242 \text{ miles/hour}\,.$$

3. Let x be the amount of water to be added. The equation that gives the amount of pure nitric acid is:
Part (a)

$$(0.6)(26) = (0.4)(26 + x)$$
$$156 = 104 + 4x$$
$$x = 52/4 = 13 \text{ gallons}\,.$$

Part (b)

$$(0.6)(26) = (0.2)(26 + x)$$
$$156 = 52 + 2x$$
$$x = 104/2 = 52 \text{ gallons}\,.$$

Part (c) No.

4. Let x = number of gallons at the start. Then we have $x + 9000$ gallons at the end,

$$55x = 46(x + 9000)\,, \quad 55x - 46x = 414{,}000 \quad x = 414{,}000/9 = 46{,}000 \text{ gallons}\,.$$

5. Let
r_1 = Arvin's rate = 1/6 total lawn/hour,
r_2 = Bozo's rate = 1/2 total lawn/hour. Then
$t(1/6 + 1/2) = 1$ (time required to mow the entire yard.
$t(4/6) = 1$, then $t = 6/4 = 3/2 = 1.5 = 1$ hour, 30 minutes.

6. Let the rate of painting the room (for Alice and Janet) be

$$r_1 = 1/(5.5) = 2/11\,, \quad r_2 = 1/(4.5) = 2/9\,.$$

To paint the entire room, we have $t[2/(11) + 2/9] = 1$.

$$r_1 + r_2 = \frac{2}{11} + \frac{2}{9} = \frac{2(9+11)}{(9)(11)} = \frac{40}{99}.$$

So, $t(40/99) = 1$ for the entire room. Thus, $t = 99/40 = 2.475$ hour or 2 hours and 28.5 minutes.

7. Let x be the first integer. Then

$$x + (x+1) + (x+2) = 249, \quad 3x + 3 = 249, \quad 3x = 246, \quad x = 82.$$

A clever alternate is: let x be the middle integer. Then

$$(x-1) + x + (x+1) = 249, \quad 3x = 249, \quad x = 83.$$

Then the middle integer is 83, so the first integer is 82.

Of course one can easily guess at the answers and in this way solve the problem experimentally. Remember that we are trying to learn algebra, — not guessing. Please look at problem 12.

8. Following the pattern in problem 7,

$$x + (x+1) + (x+2) + (x+3) = 274,$$

$$4x + 6 = 274, \quad 4x = 268, \quad x = 67.$$

9. Now successive integers differ by 2. As in problem 8,

$$x + (x+2) + (x+4) + (x+6) = 1096,$$

$$4x + 12 = 1096, \quad 4x = 1084, \quad x = 271.$$

10. As in problem 9 we will need the sum of *five* integers.

$$2 + 4 + 6 + 8 + 10 = 30.$$

Then the basic equation will give

$$6x + 30 = 492, \quad 6x = 462, \quad x = 77.$$

11. Let x and y be the two numbers with x the larger number. The problem states that

$$x - y = 4 \quad \text{and} \quad x^2 - y^2 = 200.$$

Of course we can guess at a solution, but we are learning algebra.

Since $x = y + 4$ we use this in the second equation and find that

$$(y+4)^2 - y^2 = 200\,.$$

If we expand the square and drop the terms $y^2 - y^2$ we will have $8y + 16 = 200$ and hence $y = 184/8 = 23$. Then $x = y + 4 = 27$. Thus $y = 23$ and $x = 27$.

12. As in problem 11 we have

$$x - y = 6\,, \quad x^2 - y^2 = 267\,.$$

So

$$(y+6)^2 - y^2 = 267\,, \quad 12y + 36 = 267\,,$$

$$y = \frac{(267-36)}{12} = \frac{231}{12} = 19\frac{1}{4} \quad \text{and} \quad x = 25\frac{1}{4}\,.$$

Notice that in this problem, the answers are *not integers.*
So if you tried to guess at the answers you would have very great difficulty.

13. Let x and y be the two numbers. Then

$$x = 6y \quad \text{and} \quad x + y = 6y + y = 513\,.$$

So,

$$7y = 513\,, \quad y = \frac{513}{7} = 73\frac{2}{7} \quad \text{and} \quad x = 6y = 6\left(73\frac{2}{7}\right) = 439\frac{5}{7}\,.$$

14. If x, y, and z are the three numbers, then

$$y = 2x\,, \quad z = 3y = 3(2x) = 6x$$

and so

$$x + y + z = x + 2x + 6x = 9x = 765\,,$$

$$x = \frac{765}{9} = 85\,, \quad y = 2(85) = 170\,, \quad \text{and} \quad z = 3(170) = 510\,.$$

15. Let $d =$ the distance from Wildwood to Lakeview. Using $d = rt$ we have

$$d = 60t \quad \text{and} \quad d = 50(t + 0.25)\,.\ (15 \text{ minutes} = 0.25 \text{ hours})$$

So

$$60t = 50(t + 0.25) = 50t + 12.5 \quad \text{or} \quad 10t = 12.5\,,$$

$t = 1.25$ or one hour and 15 minutes. Then $d = 60t = 60(1.25) = 75$ miles.

16. Let $x =$ speed of the bicycle.
Then y, the speed of the car, is $x + 25$. The time that Alice rode her bicycle before Betty caught up with her is

$$1 \text{ hr } 15 \text{ minutes} + 45 \text{ minutes} = 2 \text{ hours}.$$

Therefore, $d = 2x$ for Alice, and $d = 0.75(x + 25)$ for Betty.
Hence $2x = 0.75(x + 25)$, or $2x = 0.75x + 18.75$. Therefore $1.25x = 18.75$.
Consequently $x = 18.75/1.25 = 15$ miles/hour. Then $y = 15 + 25 = 40$ miles/hour.

17. Let r be the speed of the wind. When flying against the wind the speed is $355 - r$, and when flying with the wind the speed is $355 + r$.
Since $d = rt$ is the same in both directions we have

$$6(355 - r) = 4(355 + r) \quad \text{or}$$

$$2130 - 6r = 1420 + 4r \quad \text{or}$$

$$2130 - 1420 = 710 = 4r + 6r. \quad r = 710/10 = 71 \text{ miles/hour}.$$

18. Let $r =$ his speed on I-75. Then $r - 30$ is his speed in town.
Let d_1 and d_2 be the distances traveled in town and on I-75 respectively.
Then $d_1 + d_2 = 15$. From $d = rt$ and the statements in the problem

$$d_1 + d_2 = (r - 30)(1/4) + r(1/6) = 15,$$

$$r\left(\frac{1}{4} + \frac{1}{6}\right) - 7.5 = 15,$$

$$r\frac{5}{12} = 15 + 7.5, \quad \text{so} \quad r = 12(4.5) = 54 \text{ miles/hour}.$$

19. Let $x =$ amount invested at 6%. Then $10,000 - x$ is invested at 4%. So

$$0.06x + 0.04(10{,}000 - x) = 0.054(10{,}000).$$

Multiply both sides by 1,000 to obtain

$$60x + 40(10{,}000) - 40x = 54(10{,}000),$$

$$20x = (54 - 40)(10{,}000) = 140{,}000.$$

So $x = 140,000/20 = \$7,000$.

20. Let $x =$ the amount invested at 4%. Then

$$0.04x = 0.07(8{,}000 - x) + 155,$$

or

$$4x = 7(8{,}000) - 7x + 15{,}500,$$

or

$$4x + 7x = 56{,}000 + 15{,}500 = 71{,}500\,.$$

So $11x = 71{,}500$ or $x = 71{,}500/11 = \$6{,}500$.

21.

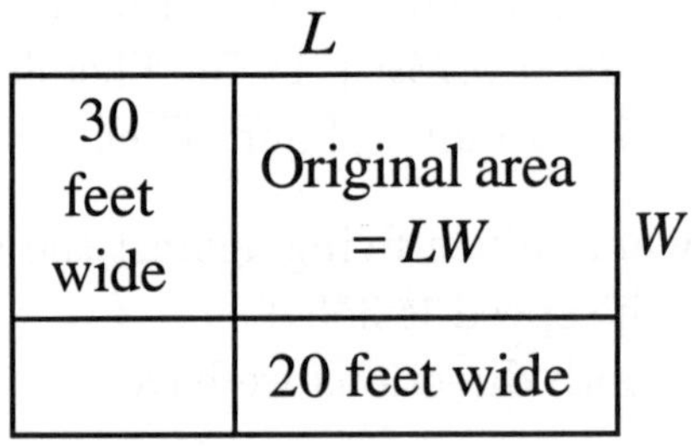

30 feet wide	Original area $= LW$
	20 feet wide

L (top), W (right side)

$30W = 20L - 700$. But $L = 2W$ so $30W = 40W - 700$, so $700 = 40W - 30W = 10W$. Thus $W = 70$ feet and $L = 140$ feet.

22. Let $x =$ the price of the mansion. Then

$$x = \frac{1}{2}x + \frac{1}{3}x + 10{,}000 + 24{,}000$$

$$x - \frac{5}{6}x = \frac{1}{6}x = 34{,}000\,.$$

Hence $x = 6(34,000) = 204,000$. The second house cost

$$\frac{204{,}000}{3} + 10{,}000 = 68{,}000 + 10{,}000 = \$78{,}000\,.$$

23. Let x, y, and z be the salaries of A, B, and C respectively. Then

$$x = 2y - 120 \quad \text{and} \quad z = \frac{2y}{3} + 90\,.$$

Then

$$x + y + z = (2y - 120) + y + \left(\frac{2y}{3} + 90\right) = 630\,.$$

Hence $11y/3 - 30 = 630$, so $y = 3(660)/11 = 180$. Thus $x = 240$ and $z = 2y/3 + 90 = 120 + 90 = \$210$. Hence Arvin at 240 has the biggest income.

24. Let $x =$ the cost of the car. Then

$$x = \left(\frac{x}{5} + 300\right) + \left(\frac{x}{5} + 300 + 90\right) + \left(\frac{x}{4} - 200\right).$$

Thus $x = x(1/5 + 1/5 + 1/4) + 490$. Thus $x - x(13/20) = 490$. Hence $x = 20(4900)/7 = \$1{,}400$.

25. Alice's rate is 1/36 of the house/hour. Cora's rate is 1/24 of the house/hour.
Let x = time Cora spent working alone. To do the whole job we have

$$1 = 12\left(\frac{1}{36} + \frac{1}{24}\right) + x\frac{1}{24}\,.$$

Thus

$$1 = \frac{1}{3} + \frac{1}{2} + \frac{x}{24}$$

$1 - 5/6 = x/24$, or $1/6 = x/24$. Hence $x = 24/6 = 4$ hours.

26. Let b and c be the rate at which Barney and Corrigan respectively can build the wall.
Working together it takes 12 hours, so $(b + c)12 = 1$.
Corrigan working alone gives $c21 = 1$, or $c = 1/21$.
If we use this in the first equation, we find that

$$\left(b + \frac{1}{21}\right) 12 = 1\,. \quad \text{Hence} \quad 12b = 1 - \frac{12}{21}\,.$$

Hence $b = (1/12) \times (9/21) = 1/28$. It would take Barney 28 hours.

27. Let d and s be the present ages of my son and daughter.
Then $d = s + 11$. In 3 years $d + 3 = 2(s + 3)$.
If we use the first equation in the second, we will obtain

$$(s + 11) + 3 = 2s + 6\,, \quad \text{or} \quad s - 2s = 6 - 14\,, \quad \text{or} \quad -s = -8\,.$$

Hence $s = 8$ and $d = 8 + 11 = 19$ years.

28. Let m and d be the ages of the mother (Arlene) and daughter respectively.
Then $m = 5d$. After 18 years $m + 18 = 2(d + 18)$. Thus

$$5d + 18 = 2(d + 18)\,, \quad \text{or} \quad 3d = 36 - 18 = 18\,.$$

Therefore $d = 6$ and $m = 5(6) = 30$ years.

29. The rates for A, B, and C are 1/36, 1/27, and 1/18 respectively.
Let t be the total number of days needed to pick the entire grove.
Since Corky starts two days later, the fraction of the grove that he picks is $(t - 2)$ $(1/18)$.
For the entire grove we have

$$1 = t\left(\frac{1}{36} + \frac{1}{27}\right) + (t - 2)\frac{1}{18}\,.$$

Now $2/18 = 1/9$, and the G.C.D. of the denominators $= 108$. Hence

$$1 + \frac{1}{9} = \frac{10}{9} = t\frac{3+4+6}{108} = t\frac{13}{108},$$

Therefore

$$t = \frac{10}{9}\frac{108}{13} = \frac{120}{13} = 9\frac{3}{13} \text{ days}.$$

30. Let $t =$ the time Arvin worked.
The rates for A, B, and C are $1/36$, $1/27$, and $1/18$ respectively.
The times worked for A, B, and C are respectively t, $(t-3)$, and $(t-6)$.
To pick the entire grove we have

$$1 = \frac{1}{36}t + \frac{1}{27}(t-3) + \frac{1}{18}(t-6)$$

As before, (problem 29) the G.C.D. $= 108$

$$1 + \frac{3}{27} + \frac{6}{18} = \frac{3+4+6}{108}t.$$

$$\frac{13}{9} = \frac{13}{108}t, \quad \text{so} \quad t = \frac{13(108)}{9(13)} = 12 \text{ days}.$$

31. Let x and y be the distances traveled by the slower and faster car respectively.
Then $y = x + 4$ when they meet for the first time after they started the race.
If t is the time elapsed then $x = t(105)$, and $y = t(120)$.
Then $y = x + 4$ gives $120t = 105t + 4$. So $15t = 4$ and $t = 4/15$ hours.
Converting this to minutes, yields $t = 4(60)/15 = 16$ minutes.

32. Let $t =$ the time measured from 4:00 P.M.
Then $65t = 75(t - 1/3)$. Thus $t = 2.5$ hrs, so the meeting time is $4 + 2.5 =$ 6:30 P.M.

33. $60t = 80(t - 1/2)$ gives $t = 2$ hrs, so the meeting time is $1 + 2 =$ 3:00 P.M.

Chapter 6, Exercise 5

1. Even. **2.** Neither. **3.** Odd. **4.** Even.

5. Even. **6.** Odd. **7.** $g(x) = 10x^2$, even.

8. $g(x) = -8x^4 + 12x^2$, even.

9. $g(x) = 2\sqrt{13}x^{10}$, even.

10. (A) $h(x) = -4x^3$, odd. (B) $h(x) = 6x^5 + 10x^3$, odd. (C) $h(x) = -2\sqrt{6}\,x^7$, odd.

11. $g(x) = f(x) + f(-x)$ is an even function.

12. $h(x) = f(x) - f(-x)$ is an odd function.

13. If we use the $g(x)$ and $h(x)$ from problems 11 and 12 we have $f(x) = (g(x) + h(x))/2$.

Chapter 7, Exercise 1

1. $-3, 1$.

2. $1, 6$.

3. $-3, 5$.

4. $-8, 3$.

5. $-6, 4$.

6. $2, 14$.

7. $-4, 9$.

8. $-4, 10$.

9. $-3, 16$.

10. $-30, -2$.

11. $3/2, 2$.

12. $-2, 9/2$.

13. $-4, 1/3$.

14. $1/2, 2/3$.

15. $-2/3, 3$.

16. $-1, 1/5$.

17. $-4/3, 3/2$.

18. $-3/2, 1/3$.

19. $-4/3, 3/2$.

20. $-1, 1/3$.

21. $-3, -2, 2, 3$.

22. $-4, -\sqrt{5}, \sqrt{5}, 4$.

23. $-\sqrt{2}, -1, 1, \sqrt{2}$.

24. $\pm\sqrt{3}, \pm\sqrt{6}$.

25. $-1, 1$.

26. $-2, 2$.

27. $-\sqrt{5}, \sqrt{5}$.

28. No real roots.

29. $x^2 - 19x + 60 = 0$.

30. $x^2 + 12\pi x + 11\pi^2 = 0$.

31. $x^2 - 2x - 48 = 0$.

32. $x^2 + 3x - 180 = 0$.

33. $x^2 + 4x - 77 = 0$.

34. $x^2 + x - 12 = 0$.

35. $6x^2 - 5x + 1 = 0$.

36. $12x^2 - x - 1 = 0$.

37. $25x^2 + 5x - 6 = 0$.

38. $6x^2 + 5x - 56 = 0$.

39. $0, -5$.

40. $0, -4$.

41. $\pm 1, \pm\sqrt{6}$.

42. $\pm 1, \pm\sqrt{5}$.

43. $3/2, 4/3$.

44. $3, 9/4$.

Chapter 7, Exercise 2

1. $-2 \pm \sqrt{2}$.

2. $\dfrac{5 \pm \sqrt{5}}{2}$.

3. $5 \pm \sqrt{15}$.

4. $-5 \pm \sqrt{26}$.

5. $\dfrac{3 \pm \sqrt{21}}{2}$.

6. $r_1 = -1$, $r_2 = 1/3$.

7. $\dfrac{4 \pm 3\sqrt{2}}{2}$.

8. $-3 \pm \sqrt{3}$.

9. $r_1 = 3/2, r_2 = 2$.

10. $-2 \pm \sqrt{10}/2$.

11. $\dfrac{5 \pm \sqrt{11}}{2}$.

12. $r_1 = 1/2, r_2 = 2/3$.

13. $-2 \pm \sqrt{7}\,i$.

14. $-1 \pm \sqrt{6}/3$.

15. $(1 \pm \sqrt{14}i)/5$.

16. $-1 \pm \sqrt{23}i$.

17. $(3 \pm \sqrt{26}i)/5$.

18. $(-5 \pm \sqrt{71}i)/2$.

19. $1, (-1 \pm \sqrt{3}i)/2$.

20. $1, (-1 \pm \sqrt{5})/2$.

21. Hint: Set $u = x^2 + 2x$ and $u + 56/u = 15$. Then $u = 7, 8$, and $x^2 + 2x = u$ (7 or 8). Solving these equations gives $x = -4, 2, -1 \pm 2\sqrt{2}$.

22. Set $u = x^2 - 3x$ and $u - 8/u = 2$. Then $u = -2$ or 4. Then the equations $x^2 - 3x = u$, where $u = -2$ or 4 give $x = -1, 1, 2, 4$.

23. $(x-1)^3 = 0$, or $x^3 - 3x^2 + 3x - 1 = 0$.

24. $(x+2)^4 = 0$, or $x^4 + 8x^3 + 24x^2 + 32x + 16 = 0$.

25. $r_1 = i, r_2 = 2 + i$.

26. $r_1 = 5i, r_2 = -i$.

27. $r_1 = 1 + i, r_2 = 2 + i$.

28. $r_1 = 6i, r_2 = -i$.

29. $(3-2i)^2 = 9 + (2)(3)(-2i) + (-2i)(-2i) = 9 - 12i - 4 = 5 - 12i$.

30. $(A + Bi)^2 = C + Di$, if and only if $A^2 - B^2 = C$ and $2AB = D$.

31. $A^2 - B^2 = 80$, and $2AB = 18$. Set $A = 9$ and $B = 1$. Ans. $\pm(9 + i)$.

32. $r_1 = -6, r_2 = 3 + i$.

33. $\pm(1 + 2i)$.

34. $\pm(-1 + 5i)$.

35. $\pm(4 + i)$.

36. $\pm(2 + 5i)$.

37. $\pm(5 + 2i)$.

38. $\pm(9 - 4i)$.

39. $r_1 \approx 1.33, r_2 \approx -4.14$.

40. $r_1 \approx 1.27, r_2 \approx -0.98$.

41. $r_1 \approx 1.68, r_2 \approx -0.66$.

42. $r_1 \approx 0.24, r_2 \approx -1.06$.

Chapter 7, Exercise 3

1. -7.

2. 8.

3. $5, 11$.

4. $-3/5, 4$.

5. $-11, 4$.

6. $2/7, 2$.

7. $2 \pm \sqrt{3}/3$.

8. $1, -8/3$.

9. $-2, 5$.

10. $-2 \pm \sqrt{3}/3$.

11. 3.

12. 9.

13. $9/2$.

14. -7.

15. -4.

16. 3.

17. $17/4$.

18. $5/4$.

19. -1.

20. $5, -2$.

21. $4, 8$.

22. $6/5, 2$.

23. $3/2, 2$.

24. $-3/2$.

25. $16/9$.

26. $1/9$.

27. $1/5, 4/5, u = 1/2, 2$.

28. $2/15, 9/35, u = 2/3, 3/2$.

29. $x = 0$ is the only solution. When $x = 5$ the equation gives $3 - 4 = -1 \neq +1$.

30. $-9/2, 2$.

31. $1, 3$.

32. 4.

33. 8.

34. 9.

Chapter 7, Exercise 4

1. x = width and y = length. Then $xy = 88$, $y = x + 3$, so $x(x+3) = 88$, $x^2 + 3x - 88 = (x+11)(x-8) = 0$, $x = -11$ is physically impossible, so $x = 8$. Width = 8 feet and length = $8 + 3 = 11$ feet.

2. $x(x+3) = 108$, so $(x+12)(x-9) = 0$. Thus $x = 9$ feet and $y = 9 + 3 = 12$ feet

3. $x(x+4) = 100$. Thus $x = -2 + \sqrt{104}$ feet and $y = x + 4 = 2 + \sqrt{104}$ feet.

4. $x = -3 + 3\sqrt{11}$, $y = 3 + 3\sqrt{11}$.

5.

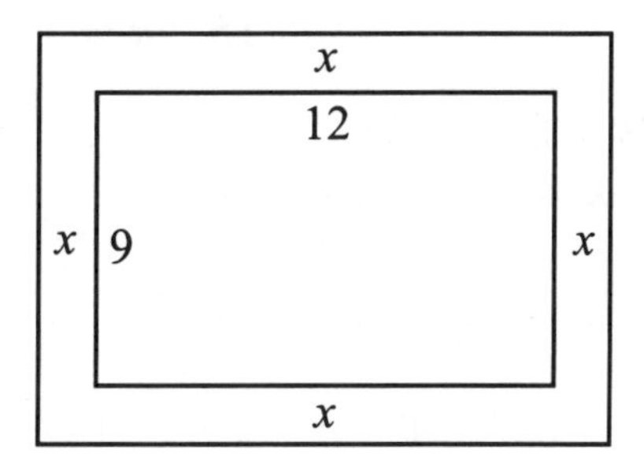

$$(9 + 2x)(12 + 2x) = 9(12) + 72 = 180\,,$$

$$108 + 24x + 18x + 4x^2 = 180\,,$$

$$4x^2 + 42x - 72 = 0\,,$$

$$2x^2 + 21x - 36 = 0\,, \quad (x + 12)(2x - 3) = 0\,.$$

The answer $x = -12$ is physically impossible. Hence width = 3/2 feet.

6. Width = 3/2 feet.

7.

$$y = x + 6\,, \quad x^2 + y^2 = 22$$

$$x^2 + (x+6)^2 = 2x^2 + 12x + 36 = 22\,,$$

$$2x^2 + 12x + 14 = 0\,, \quad x^2 + 6x + 7 = 0\,.$$

$$x = \frac{-6 \pm \sqrt{36 - 28}}{2} = \frac{-6 \pm \sqrt{8}}{2}$$

(I) $x = -3 + \sqrt{2}$, $y = 3 + \sqrt{2}$, (II) $x = -3 - \sqrt{2}$, $y = 3 - \sqrt{2}$.

8. (I) $x = (-5 + \sqrt{157})/2$, $y = x + 5$, (II) $x = (-5 - \sqrt{157})/2$, $y = x + 5$.

9. $x + y = 8$, $xy = 13$, $x(8 - x) = 13$, $-x^2 + 8x - 13 = 0$. Multiply by -1 and solve.

$$x = 4 + \sqrt{3}\,, \quad y = 4 - \sqrt{3}\,,$$

10. $x = 2, y = 7$.

11. $x, y = 5 \pm \sqrt{10}$.

12. $x, y = 15 \pm 3\sqrt{3}\,$.

13. We have a repeated root when $D = 0$.

$$D = b^2 - 4ac = 4(k^2 + 6k + 9) - 4(k^2 + 3) = 0\,,$$

$4(6k + 6) = 0$ so $k = -1$. For $k = -1$ the equation becomes $x^2 - 4x + 4 = 0$. Hence the repeated root $r = 2$.

14. $k = -3, r = 0$.

15. $k = 1/4$, $r = -1/4$.

16. $D = -3k^2 - 4k + 4 = -(3k - 2)(k + 2)$.
(I) $k = 2/3$, $r = 2/3$, (II) $k = -2$, $r = 2$.

17. (I) $k = 9, r = 6$, (II) $k = 1, r = 2$.

18. $D = (k + 2)^2 - 4k(k + 1) = -3k^2 + 4 = 0 \quad$ at $k = \pm 2/\sqrt{3}\,$.
At $k = 2/\sqrt{3}$, the equation gives

$$\frac{2}{\sqrt{3}}\left(x + \frac{1 + \sqrt{3}}{2}\right)^2 = 0\,.$$

At $k = -2/\sqrt{3}$, the equation gives

$$-\frac{2}{\sqrt{3}}\left(x + \frac{1 - \sqrt{3}}{2}\right)^2 = 0\,.$$

(I) $k = 2/\sqrt{3}\,, r = -(1 + \sqrt{3}\,)/2$, (II) $k = -2\sqrt{3}\,, r = -(1 - \sqrt{3}\,)/2$,

19. For this equation

$$D = 4(k^2 + 2kb + b^2) - 4k^2 = 4(2kb + b^2) = 0\,,$$

so $k = -b/2$. Then the equation becomes

$$x^2 + 2\left(-\frac{b}{2} + b\right)x + \left(-\frac{b}{2}\right)^2$$
$$= x^2 + bx + \frac{b^2}{4} = \left(x + \frac{b}{2}\right)^2 = 0\,.$$

So $k = -b/2$ and $r = -b/2$.

20.

$$x - y = 1 \quad \text{so } x = y + 1\,,$$

$$x^3 - y^3 = (y+1)^3 - y^3 = y^3 + 3y^2 + 3y + 1 - y^3 = 10\,,$$

$$3y^2 + 3y - 9 = 3(y^2 + y - 3) = 0\,. \quad y = (-1 \pm \sqrt{13}\,)/2$$

(I) $y = (-1 + \sqrt{13}\,)/2$, $x = (1 + \sqrt{13}\,)/2$, (II) $y = (-1 - \sqrt{13}\,)/2$, $x = (1 - \sqrt{13}\,)/2$,

21.

$$x = y + 2\,, \quad (y+2)^3 - y^3 = 98\,.$$

$$y^3 + 6y^2 + 12y + 8 - y^3 = 98\,,$$

$$0 = 6y^2 + 12y - 90 = 6(y^2 + 2y - 15) = 6(y+5)(y-3)$$

Hence $y = -5$ or $y = 3$. Thus (I) $x = -3,\ y = -5$, (II) $x = 5,\ y = 3$.

22. As in problem 21, $6y^2 + 12y + 8 - 32 = 0$ or $6(y^2 + 2y - 4) = 0$, etc.

(I) $x = 1 + \sqrt{5}$, $y = -1 + \sqrt{5}$, (II) $x = 1 - \sqrt{5}$, $y = -1 - \sqrt{5}$.

23.

$$x - y = -1\,, \quad x = y - 1\,.$$

$$x^3 - y^3 = (y-1)^3 - y^3 = 8\,,$$

$$-3y^2 + 3y - 1 - 8 = 0\,,$$

$$-3(y^2 - y + 3) = 0\,.$$

$$y = (\pm\sqrt{1-12}\,)/2 = (1 \pm i\sqrt{11}\,)/2\,,$$

because $\sqrt{-11} = \sqrt{(-1)(11)} = i\sqrt{11}$. $x = y - 1 = (-1 \pm i\sqrt{11}\,)/2$.

(I) $x = (-1 + i\sqrt{11})/2$, $y = (1 + i\sqrt{11})/2$, (II) $x = (-1 - i\sqrt{11})/2$, $y = (1 - i\sqrt{11})/2$,

24. For (I) let $x = (-1 + i\sqrt{11}\,)/2$, then $y = x + 1$

$$x^3 = \frac{1}{8}(-1 + 3i\sqrt{11} - 3(-11) - i11\sqrt{11}\,)\,,$$

$$y^3 = \frac{1}{8}(1 + 3i\sqrt{11} - 3(11) - i11\sqrt{11}\,)\,,$$

$$x^3 - y^3 = \frac{1}{8}(-1 - 1 + 33 + 33)$$

$$= 8\,.$$

The check for (II) is similar.

Chapter 8, Exercise 1

1. ABC, ACB, BAC, BCA, CAB, CBA.

2.
$AB + BC = AC$, $\quad 5 + 8 = 13.$
$AC + CB = AB$, $\quad 13 - 8 = 5,$
$BA + AC = BC$, $\quad -5 + 13 = 8,$
$BC + CA = BA$, $\quad 8 - 13 = -5,$
$CA + AB = CB$, $\quad -13 + 5 = -8,$
$CB + BA = CA$, $\quad -8 - 5 = -13.$

3. $r = \sqrt{x_2^2 + y_2^2} = \sqrt{x^2 + y^2}$, when P_2 is $P(x, y)$.

4. $y = -3$. **5.** $x = -5$. **6.** $y = 5$. **7.** $x = 2\sqrt{3}$.

8. A horizontal line 6 units above the x-axis.

9. A vertical line 3 units to the left of the y-axis.

10. A line through the origin that makes an angle of 45 degrees with the x-axis.

11. A line through the origin that makes an angle of 135 degrees with the x-axis.

12. The line of problem 10 moved upward 1 unit.

13. A circle with center at the origin and radius 7 units.

14. The two lines of Problems 10 and 11. This is called the union of those two lines.

15. The same as in Problem 10.

16. The same as in Problem 14.

17. 13. **18.** 5. **19.** $2\sqrt{5}$. **20.** $16\sqrt{2}$.

21. $y = x - 4$. Hint: Use $(x + 3)^2 + (y - 3)^2 = (x - 7)^2 + (y + 7)^2$.

22. $x + 2y = 3$. Hint: Use $(x + 1)^2 + (y + 3)^2 = (x - 3)^2 + (y - 5)^2$.

23. $9x + y = 10$. Hint: Use $(x + 4)^2 + (y - 5)^2 = (x - 5)^2 + (y - 6)^2$.

24. $8x - 18y = 69$. Hint Use: $(x - 1)^2 + (y - 2)^2 = (x - 5)^2 + (y + 7)^2$.

25. $x^2 + y^2 = 6x - 8y$. Hint: Use $(x - 3)^2 + (y + 4)^2 = 5^2$.

26. $x^2 + y^2 + 8x - 10y = 40$.

27. $x^2 + y^2 - 6x - 4y + 12 = 0$.

28. $x^2 + y^2 - 8x + 2y = 104$.

29. $A(1, 3)$, $B(4, 2)$, $C(-2, -6)$.
$|AB|^2 = 10$, $|AC|^2 = 90$, $|BC|^2 = 100$. Yes. It is a right triangle.

30. $A(8,-3)$, $B(-4,2)$, $C(1,6)$.
$|AC|^2 = 130$, $|BC|^2 = 41$, $|AB|^2 = 169$. No. It is not a right triangle.

31. $A(-3,1)$, $B(1,-4)$, $C(11,4)$.
$|AC|^2 = 205$, $|BC|^2 = 164$, $|AB|^2 = 41$. Yes. It is a right triangle.

32. $A(-4,2)$, $B(2,-1)$, $C(0,-5)$.
$|AC|^2 = 65$, $|BC|^2 = 20$, $|AB|^2 = 45$. Yes. It is a right triangle.

33. $A(2,-6)$, $B(-5,1)$, $C(-1,5)$.
$|AC|^2 = 130$, $|BC|^2 = 32$, $|AB|^2 = 98$. Yes. It is a right triangle.

34. $A(5,-2)$, $B(6,5)$, $C(2,2)$.
$|AC|^2 = 25$, $|BC|^2 = 25$. The triangle is isosceles.

35. $A(2,1)$, $B(9,3)$, $C(4,-6)$.
$|AC|^2 = 53$, $|AB|^2 = 53$. The triangle is isosceles.

Chapter 8, Exercise 2

1. (I) The modern cash register does subtractions automatically. But children should learn and practice subtraction in school.
(II) First typewriters, and now printers and word processors make handwriting obsolete. But children should learn and practice writing "by hand" in school.

Chapter 8, Exercise 3

1. $(0,5)$, $(-3,-1)$.

2. $(0,7)$, $(3,-2)$.

3. $(0,-4)$, $(4,0)$.

4. $(0,-5)$, $(-5,5)$.

5. $(0,2)$, $(-6,0)$.

6. $(0,-2)$, $(-8,0)$.

7. $(0,-4)$, $(-10,0)$.

8. $(0,-8)$, $(6,0)$.

9. Some points on the graph are $(0,0)$, $(\pm1,1)$, $(\pm2,4)$, $(\pm3,9)$, $(\pm4,16)$.

10. The graph of Problem 9 turned upside down.

11. The graph of Problem 9 shrunk in the direction of the y-axis by the factor $1/4$.

12. The graph of Problem 9 moved downward 3 units.

13. The graph of Problem 9 moved to the right 3 units.

14. The graph of Problem 9 moved to the left 3 units.

15. The graph of Problem 13 moved upward 4 units.

16. This graph is the same as the one in Problem 15.

17. This graph is the same as the one in Problem 13.

18. $x^2 + y^2 - 2x + 4y - 4 = 0.$

19. $x^2 + y^2 - 8x + 6y = 0.$

20. $x^2 + y^2 - 2x + 4y - 95 = 0.$

21. $x^2 + y^2 - 4x + 6y - 23 = 0.$

22. A circle center at $(2, -1)$, radius 5.

23. A circle center at $(-3, -4)$, radius 1.

24. The point $(3, 8)$.

25. No points in the plane.

26. $(x - 3)^2 + (y + 4)^2 = 9$, center $(3, -4)$ radius 3.

27. $(x - 8)^2 + (y - 3)^2 = 9$, center $(8, 3)$ radius 3.

28. A wedge with vertex at the origin. Two half lines which meet at the origin. One makes an angle of 45° with the x-axis and the other makes an angle of 135° with the x-axis.

29. The graph of Problem 28 moved to the right 5 units.

30. The graph of Problem 28 moved to the left 6 units.
31. The graph of Problem 28 moved downward 5 units.

32. The graph of Problem 9 rotated clockwise 90°, plus its reflection in the y-axis.

33. The graph of Problem 28 rotate clockwise 90°.

Chapter 8, Exercise 4

1. $m = \dfrac{-7-3}{4-2} = \dfrac{-10}{2} = -5.$

2. $-5.$

3. $\dfrac{13}{6}.$

4. $\dfrac{1}{3}.$

5. 1.

6. $\sqrt{3}\,.$

7. Undefined or ∞.

8. $\dfrac{8/3 - 3/4}{10/3 + 7/2} = \dfrac{(32-9)/12}{(20+21)/6} = \dfrac{23}{82}.$

9. $(y - y_1)/(x - x_1) = m$, gives $(y - 3)/(x - 2) = -5$, or

$$y - 3 = -5(x - 2)\,, \quad \text{or } y - 3 = -5x + 10\,, \quad \text{or}$$
$$y = -5x + 10 + 3\,, \quad \text{or } y = -5x + 13\,.$$

10. $y = -5x - 23.$

11. $y = \dfrac{13x}{6} - \dfrac{23}{6}\,.$

12. $y = x/3 - 13/3.$

13. $y = x + 14.$

14. $y = \sqrt{3}x - \sqrt{3}\,.$

15. This is a vertical line $x = -3$, so the form $y = mx + b$ is impossible.

16. $(y - y_1)/(x - x_1) = m$, gives $(y - 3/4)/(x + 7/2) = 23/82$, $y - 3/4 = (x + 7/2)23/82$,

$$y = \frac{23}{82}x + \frac{23}{82}\frac{7}{2} + \frac{3}{4}, \qquad y = \frac{23}{82}x + \frac{23}{41}\frac{7}{4} + \frac{41(3)}{41(4)},$$

$$y = \frac{23}{82}x + \frac{161}{41(4)} + \frac{123}{41(4)}, \quad y = \frac{23}{82}x + \frac{71}{41}.$$

17. For Problem 9, the equation for the straight line is $y = -5x + 13$. As a check we have: For the point $(2, 3)$, $3 = -5(2) + 13$, For the point $(4, -7)$, $-7 = -5(4) + 13$. The other problems, 10 through 16 are treated similarly.

18. $m = -3/5$, $b = 7$.

19. $m = -5/3$, $b = -12$.

20. $m = -1$, $b = 2$.

21. $m = 7/3$, $b = 19$.

22. $m = -1/2$, $b = 3/2\pi$.

23. $m = -7/9$, $b = -20/3$.

24. From the equation, if $y = 0$, then $x = a$, and if $x = 0$, then $y = b$.

25. $4x - 5y = 20$.

26. $2x + y = 2$.

27. $2y - 7x = 14$.

28. $-2x + 6y = 1$.

29. The two lines $y = 2x$, and $y = -2x$.

30. The two lines $y = x + 4$, and $y = -x + 2$.

31. The two lines $y = x + 1$, and $y = -x + 3$.

32. The half lines $y = 1 + 2x$, if $x \geq 0$; and $y = 1$, if $x < 0$.

33. The half lines $y = 2 + x$, if $x \geq 0$; and $y = 2 - x$, if $x < 0$.

34. The half lines $y = 2 - x$, if $x \geq 0$; and $y = 2 + x$, if $x < 0$.

35. The half lines $y = 1 + x$, if $x \geq 2$; and $y = 5 - x$, if $x < 2$.

37. Yes, $m = -1/2$.

38. Yes, $m = 2/5$.

39. No, $-3/5 \neq -11/18$.

40. No, $8/5 \neq 13/8$.

Chapter 8, Exercise 5

1. $y = x - 10$.

2. $2y = x + 22$.

3. $3y = -x$.

4. $5y = 2x - 3$.

5. $x = 100$.

6. $28x - 24y = 1$.

7. $y = -x$.

8. $y = -2x + 11$.

9. $y = 3x$.

10. $2y = -5x - 7$.

11. $y = 200$.

12. $18x + 21y = 31$.

13. $(1,-2)$. **14.** $(2,-1)$. **15.** $(2,-3)$.

16. $(4,3)$. **17.** $(1,2)$.

18. For the line segment AP the slope is $m_1 = y/(x+r)$ and for BP the slope is $m_2 = y/(x-r)$. Then $m_1m_2 = y^2/(x^2-r^2) = y^2/(-y^2) = -1$.

19. $M(8,15)$. **20.** $M(6,13)$. **21.** $M(-2,-4)$.

22. $M(-5,4)$. **23.** $M(5/12,1/2)$. **24.** $M(-7/2,-3/2)$.

25. $|P_1P_M|^2 = |P_MP_2|^2 = \left(\frac{x_2-x_1}{2}\right)^2 + \left(\frac{y_2-y_1}{2}\right)^2$, slope P_1M = slope $P_1P_2 = \frac{y_2-y_1}{x_2-x_1}$.

Chapter 9, Exercise 1

1. (A) $[-3,7]$, (B) $[2,4]$.

2. (A) $[-10,-2]$, (B) $[-6,-5]$.

3. (A) $(-3,7)$, (B) $(2,4)$.

4. (A) $(-3,7)$, (B) $[2,4]$.

5. (A) $[-3,7]$, (B) $(2,4)$.

6. (A) $[-10,-2]$, (B) $(-6,-5)$.

7. (A) $(-10,-2)$, (B) $[-6,-5)$.

8. (A) $(-7,5)$, (B) $(1,2)$.

9. (A) $(-7,5)$, (B) $[1,2]$.

10. (A) $(-7,5)$ (B) $(1,2]$.

11. (A) $[-7,5)$, (B) $[1,2)$.

12. (A) the two intervals $[1,2]$ and $[3,5]$. We have no notation for this type of set other than $[1,2] \cup [3,5]$. (B) ϕ the empty set.

13. (A) the set of all integers. (B) ϕ, the empty set.

14. $X = Y$ if every element of X is an element of Y, and every element of Y is an element of X.

15. Yes.

16. Yes

17. Yes, yes. (A) $X \cup Y = Y \cup X$, and (B) $X \cap Y = Y \cap X$.

18. Yes. Without this rule mathematicians would not be able to prove anything.

19. Yes.

20. (A) It is not commutative since $X \subset Y$ does not imply that $Y \subset X$. (B) The symbol is transitive, since $X \subset Y$ and $Y \subset Z$ implies that $X \subset Z$.

Chapter 9, Exercise 2

1. $x > y$. Square the expression for x and the one for y.

2. $x > y$. Consider $4 = \sqrt{16}$ and prove that $\sqrt{16} + \sqrt{15} > \sqrt{17} + \sqrt{14}$.

3. $y > x$. Square both terms.

4. $x < y$. Raise both terms to the sixth power, $512 < 529$.

5. If we square both terms this will lead to $A^2 - B^2 < A^2 - 1$.

6. If we replace a by $-a$, the right side is unchanged, but the left side is either unchanged or decreases. The same is true if b, c, d, are replaced by $-b, -c$, and $-d$ respectively.

7. Start with $(a-1)^2 \geq 0$. Equality if and only if $a = 1$.

8. Start with $(2a - 5b)^2 \geq 0$. Equality if and only if $2a = 5b$.

9. Use Problem 7 with $a = \sqrt{c/d}$.

10. Start with $(c-d)^2 \geq 0$ and add $4cd$ to both sides. Equality if and only if $c = d$.

11. For the left inequality start with $(\sqrt{a} - \sqrt{b})^2 \geq 0$. Equality if and only if $a = b$. For the right inequality replace a and b by $1/a$ and $1/b$. Then take reciprocals on both sides using Theorem 7, and simplify the result. This inequality will be generalized in the next section.

12. Expansion of the left side gives $a^2+7ab+10b^2 \geq 9ab+9b^2$ and this leads to $a^2+b^2 \geq 2ab$, see Example 1.

13. $(x-2y)^2 \geq 0$, with equality if and only if $x = 2y$. Or use Example 1 with $a = x$, and $b = 2y$.

14. $(x-y)^2 \geq 0$, $(y-z)^2 \geq 0$ and $(z-x)^2 \geq 0$. Then add these inequalities and divide by 2. Equality if and only if $x = y = z$.

15. Multiply both sides by c^2d^2 to obtain

$$c^4 + d^4 + 6c^2d^2 \geq 4c^3d + 4cd^3\,.$$

Put all of the terms on the left side and use $(c-d)^4 \geq 0$. This inequality is equivalent to $(c-d)^4 \geq 0$. The method does not work for $(c-d)^3$ unless we assume that $c \geq d$.

16. $(a+3b)^2 \geq 12ab$, or $(a-3b)^2 \geq 0$. Equality if and only if $a = 3b$.

17. This is equivalent to $(c^2 - d^2)(c-d) \geq 0$. If $c > d$, then both factors are positive. If $c < d$, then both factors are negative and the product is positive. Equality if and only if $c = d$.

18. As in example 1, $2\sqrt{AB}\,\sqrt{CD} \leq AB + CD$ and $2\sqrt{AC}\,\sqrt{BD} \leq AC + BD$. Then the product of these two inequalities will give the result. Equality if and only if $A = D$, AND $B = C$.

19. Problems 10, 12, 13, 14, and 15 are meaningful and true when some of the variables are negative.

20. A, B, and C are true.

21. First prove that $n \le k(n-k+1)$ for each k from 1 to n. This is equivalent to the inequality $k(k-1) \le n(k-1)$. Now take the product of all the inequalities,

$$n \le k(n-k+1) \quad \text{for} \quad k = 1, 2, 3, \ldots, n\,.$$

22. Use Theorem 9 with $n = p$, and Theorem 10 with $n = q$.

Chapter 9, Exercise 3

1. $\sqrt{4} = 2$.
2. $\sqrt{4} = 2$.
3. $\sqrt{30}$.
4. $\sqrt{54} = 3\sqrt{6}$.
5. $\sqrt{66}$.
6. $\sqrt{66}$.
7. $2\sqrt{2}$.
8. $\sqrt{30}$.
9. $2\sqrt{6}$.
10. $4\sqrt{13}$.
11. 2.
12. $2\sqrt{3}$.
13. 0.
14. $23/3\sqrt{86}$.
15. $\sqrt{5}/3$.
16. $-19/33$.
17. $181/3\sqrt{3685}$.
18. $5/\sqrt{33}$.
19. Yes.
20. (I) 16, (II) 32, (III) 2^n.

21. Yes. Take $T : (y_1 - x_1, y_2 - x_2, \ldots, y_n - x_n)$.

23. Here is the correct definition. Two points P_j and P_k are adjacent in a unit cube in $E^{(n)}$ if their coordinates differ in exactly one place. For example the two vertices $(0,1,1,0,1,1)$ and $(0,1,0,0,1,1)$ are adjacent in $E^{(6)}$.

24. Here is the correct definition. A diagonal of the unit cube in $E^{(n)}$ is an edge that joins two NONADJACENT vertices.

25. $\sqrt{3}$.
26. (I) 2, (II) $\sqrt{5}$, (III) $\sqrt{n}$.

Chapter 9, Exercise 4

1. $H = G = A = 7$.

2. $H = 4/3 \approx 1.333$, $G = \sqrt[4]{4} \approx 1.414$, $A = 1.5$.

3. $H = 4/1.655 \approx 2.417$, $G = \sqrt[4]{120} \approx 3.310$, $A = 17/4 = 4.25$.

4. $H = 4/1.45 \approx 2.759$, $G = \sqrt[4]{80} \approx 2.991$, $A = 13/4 = 3.25$.

5. $H = 10/3 \approx 3.333$, $G = \sqrt[5]{432} \approx 3.366$, $A = 17/5 = 3.4$.

6. $H = 5/1.925 \approx 2.597$, $G = \sqrt[5]{800} \approx 3.807$, $A = 26/5 = 5.2$.

7. If negative numbers are permitted to enter, then A could be 0 or negative while G could be positive, and $G > A$. If 0 can be a number in the sequence, then H is not well-defined but one can argue that in this case $H = 0$.

8. $|AB| = \sqrt{8} = 2\sqrt{2}$, $|OA| = \sqrt{4} = 2$, $|OB| = \sqrt{4} = 2$.

9. $|AB| = \sqrt{30}$, $|OA| = \sqrt{86}$, $|OB| = \sqrt{36} = 6$.

10. $|AB| = \sqrt{24} = 2\sqrt{6}$, $|OA| = \sqrt{30}$, $|OB| = \sqrt{54} = 3\sqrt{6}$.

11. $|AB| = \sqrt{208} \approx 14.42$, $|OA| = \sqrt{66}$, $|OB| = \sqrt{66} \approx 8.12$.

12. $|AB| = \sqrt{4} = 2$, $|OA| = \sqrt{165}$, $|OB| = \sqrt{201}$.

13. $|AB| = \sqrt{12} = 2\sqrt{3}$, $|OA| = \sqrt{9} = 3$, $|OB| = \sqrt{33}$.

14. Equality holds in the triangle inequality if and only if the three points A, O, and B are on a line with O on A or B or between A and B.

15. $L = 8$, $R = \sqrt{4}\sqrt{16} = 8$.

16. $L = 68$, $R = \sqrt{86}\sqrt{54} = 6\sqrt{129} \approx 68.147$.

17. $L = 50$, $R = \sqrt{30}\sqrt{86} = \sqrt{2580} \approx 50.794$.

18. $L = 60$, $R = \sqrt{59}\sqrt{66} = \sqrt{3894} \approx 62.402$.

19. $L = 190$, $R = \sqrt{165}\sqrt{220} = \sqrt{36{,}300} \approx 190.526$.

20. $L = 39$, $R = \sqrt{28}\sqrt{55} = \sqrt{1540} \approx 39.242$.

24. Yes. In the "proof" of $G \leq A$, we created a sequence $[B_k^*]$ such that $G_1 \leq G_2 \leq G_3 \leq \cdots G_{n'}$, and $A_1 = A_2 = A_3 = \cdots = A_n = A$. But we never proved that $G_k \leq A_k$. The feeling is that G_k approaches A from below, but to complete the proof we need some results about "limits" that are not properly in algebra. Consequently, we gave a persuasive argument that should satisfy the student.

Chapter 9, Exercise 5

1. $x > 11$, or $(11, \infty)$.

2. $x < 14$, or $(-\infty, 14)$.

3. $x \geq -17/4$, or $[-17/4, \infty)$.

4. $x \leq 2/3$, or $(-\infty, 2/3]$.

5. $x \geq 3$, or $[3, \infty)$.

6. $x > -9$, or $(-9, \infty)$.

7. $(2/7, \infty)$.

8. $(6, \infty)$.

9. $(-\infty, 59/3]$.

10. $(-\infty, -4/3)$.

11. $(-2/3, 3)$.

12. $(-\infty, 8/5) \cup [11/2, \infty)$.

13. $(-\infty, -\sqrt{10}] \cup [\sqrt{10}, \infty)$.

14. $(-\infty, -5/4) \cup (3/2, \infty)$.

15. ϕ.

16. $[(-1 - \sqrt{5})/2, (-1 + \sqrt{5})/2]$.

17. $(-\infty, -3] \cup [2, \infty)$.

18. $(-2, 12)$.

19. $[-6,-1]$.

20. $(-\infty,-2)\cup(14/3,\infty)$.

21. $(-7/2,9)$.

22. $(-\infty,-18]\cup[10,\infty)$.

23. $[-5/2,21/4]$.

24. $(-1,0)\cup(4,\infty)$.

25. $(-3,0)\cup(5,\infty)$.

26. $(-\infty,-12)\cup(-7,5)$.

27. $(-1/3,1)\cup(4/3,\infty)$.

28. $(-\infty,-4)\cup(-3,-2)\cup(-1,1)\cup(2,3)\cup(4,\infty)$.

29. $(-\infty,-4)\cup(-\sqrt{5.5}\,,-2)\cup(2,\sqrt{5.5}\,)\cup(4,\infty)$.

30. $(-\infty,-7]\cup[-3,\infty)$.

31. $(-\infty,-4)\cup(1,\infty)$.

32. $[-7,-1]\cup[1,\infty)$.

33. $(-\infty,-2]\cup[0,2]$.

Chapter 10, Exercise 1

1. Yes.

2. Yes. Use a plane that passes through the vertex of the cone, but contains no other point of the cone.

3. Yes. Use a plane that passes through the vertex of the cone and two distinct points of the base circle.

4. No.

5. Yes. Use a plane that passes through the vertex of the cone and only one point of the base circle

Chapter 10, Exercise 2

1. $(0,1)$, $y=1$.

2. $(0,1/4)$, $y=-1/4$.

3. $(0,1/16)$, $y=-1/16$.

4. $(0,8)$, $y=-8$.

5. Interchange x and y in Figure 2 and in the proof of Theorem 1 as given in the text.

6. Replace p by $-p\,(p>0)$ in Figure 2 and in the proof of Theorem 1.

7. $(1,0)$, $x=-1$.

8. $(0,-1)$, $y=1$.

9. $(-1,0)$, $x=1$.

10. $(0,-2)$, $y=2$.

11. $(-1/4,0)$, $x=1/4$.

12. $(0,-7/20)$, $y=7/20$.

14. $8y=x^2-8x+40$.

15. $4x=-(y^2+14y+53)$.

16. If we complete the square in $y = Ax^2 + Bx + C$ we have

$$\frac{y}{A} = \left(x^2 + \frac{B}{A}x + \frac{B^2}{4A^2}\right) + \frac{C}{A} - \frac{B^2}{4A^2}$$

$$= \left(x + \frac{B}{2A}\right)^2 + \frac{C}{A} - \frac{B^2}{4A^2}$$

$$= \left(x + \frac{B}{2A}\right)^2 + \frac{4AC - B^2}{4A^2}.$$

Thus

$$\frac{y}{A} - \frac{4AC - B^2}{4A^2} = \left(x + \frac{B}{2A}\right)^2,$$

or, finally

$$y - \frac{4AC - B^2}{4A} = A\left(x + \frac{B}{2A}\right)^2.$$

From this equation we see that $4p = 1/A$ or $p = 1/4A$. Further the vertex is

$$\left(-\frac{B}{2A}, \frac{4AC - B^2}{4A}\right) = (x_v, y_v).$$

To obtain the focus add $p = 1/4A$ to y_v. The equation of the directrix is $y = y_v - p$.

17. First divide both sides by 8. Then $A = 1/8$, $B = -1$, and $C = 5$. This will give $p = 2$, the vertex is at $(4, 3)$ and the two other items are obtained by adding and subtracting $p = 2$. Thus the focus is at $(4, 5)$ and the directrix is the line $y = 1$. Compare this result with that of problem 14.

18. First divide both sides by 2. Then $A = 3/2$, $B = 2$, and $C = -5/2$. Next $p = 1/4A = 1/6$. The formula for the vertex gives $(-2/3, -19/6)$. To obtain the other items add and subtract $p = 1/6$. Focus $(-2/3, -3)$, directrix $y = -10/3$.

19. This parabola opens downward. $p = 1/4A = -1/4 = -0.25$. The vertex is $(-2, 6)$, the focus is $(-2, 5.75)$, the directrix is $y = 6.25$.

20. Use congruent triangles to prove that if PQ is bisected by the axis and P is a point of the parabola, then Q is a point of the parabola.

Chapter 10, Exercise 3

1. $2a$,

2. $2b$.

3. Use congruent triangles.

4. Same method as in Problem 3.

5. From $b^2 = a^2 - c^2$, $b^2 = 25 - 9 = 16$. Equation:

$$\frac{x^2}{25} + \frac{y^2}{16} = 1.$$

6. As in Problem 5, $b^2 = 100 - 36 = 64 = 8^2$, so $\dfrac{x^2}{100} + \dfrac{y^2}{64} = 1.$

7. As in Problem 5, $b^2 = 25 - 24 = 1$, so $\dfrac{x^2}{25} + \dfrac{y^2}{1} = 1.$

8. As in Problem 5, $b^2 = 25 - 1 = 24$, so $\dfrac{x^2}{25} + \dfrac{y^2}{24} = 1.$

9. $b^2 = a^2 - c^2$ or $36 = a^2 - 25$. Then $a^2 = 61$, so $\dfrac{x^2}{61} + \dfrac{y^2}{36} = 1.$

10. $b^2 = 81 - 64 = 17$, so $\dfrac{x^2}{81} + \dfrac{y^2}{17} = 1.$

11. No. If yes we would have $4x^2 = 9x^2 - x^2$ or $4 = 9 - 1$.

12. $c^2 = a^2 - b^2$, so $c^2 = 26 - 12 = 14$. Foci at $(\pm\sqrt{14}, 0)$

13. $c^2 = 4 - 3 = 1$. Foci at $(\pm 1, 0)$.

14. $c^2 = 4 - 1/2 = 7/2$. Foci at $(\pm\sqrt{7/2}, 0)$.

15. Substitute the coordinates into Equation (22) and solve the pair of equations for a^2 and b^2. This will give $a^2 = 20$, $b^2 = 5$.

$$\frac{x^2}{20} + \frac{y^2}{5} = 1.$$

16. $a^2 = 30$, and $b^2 = 10$. $\dfrac{x^2}{30} + \dfrac{y^2}{10} = 1.$

17. The form is the same as Equation (22) except that now we have $b > a > 0$, and $c^2 = b^2 - a^2$.

18. Yes, set $a = b$.

19. $\dfrac{(x-h)^2}{a^2} + \dfrac{(y-k)^2}{b^2} = 1.$

In (I) we must have $a > b > 0$. In (II) we must have $b > a > 0$.

20. $\dfrac{(x-5)^2}{25} + \dfrac{(y-3)^2}{9} = 1.$

(a) the foci are $(1, 3)$, $(9, 3)$, (b) the vertices are $(0, 3)$, $(10, 3)$, (c) the axes of symmetry are $x = 5$, $y = 3$.

21. $\dfrac{(x-1)^2}{1/4}+\dfrac{(y+2)^2}{4}=1.$

(a) $(1, -2\pm\sqrt{15}/2)$, (b) $(1,0)$, $(1,-4)$, (c) $x=1$, $y=-2$.

22. $\dfrac{(x+2)^2}{16}+\dfrac{(y-6)^2}{25}=1.$

(a) $(-2,9)$, $(-2,3)$, (b) $(-2,11)$, $(-2,1)$, (c) $x=-2$, $y=6$.

25. All of the ellipses have the same foci $(0,\pm2)$, and the coordinate axes are axes of symmetry.

Chapter 10, Exercise 4

1. $2a$

2. Use congruent triangles.

3. Same method as in Problem 2.

4. In the equation $x^2/a^2 - y^2/b^2$ we have $a=8$, $b=6$.

5. $a=16$, $b=12$.

6. $a=5$, $b=\sqrt{11}$.

7. $(\pm5,0)$.

8. $(\pm3,0)$.

9. $(\pm3,0)$.

10. $(\pm5/\sqrt{6},0)$.

11. (a) $y=\pm4x/3$, (b) $y=\pm\sqrt{7}\,x/\sqrt{2}$, (c) $y=\pm\sqrt{2}\,x/2$, (d) $y=\pm\sqrt{2}\,x/\sqrt{3}$.

12. $3x^2-y^2=3$.

13. $y^2/a^2-x^2/b^2=1$.

14. (a) $\dfrac{(x-h)^2}{a^2}-\dfrac{(y-k)^2}{b^2}=1$, (b) $\dfrac{(y-k)^2}{a^2}-\dfrac{(x-h)^2}{b^2}=1.$

15. (a) $y=\pm\dfrac{b}{a}(x-h)+k$, (b) $y=\pm\dfrac{a}{b}(x-h)+k.$

16. (a) $(-3,7)$, $(-3,-3)$, (b) $(-3,5)$, $(-3,-1)$, (c) $x=-3$, $y=2$.

17. (a) $(-2,5\pm\sqrt{2})$, (b) $(-2,6)$, $(-2,4)$, (c) $x=-2$, $y=5$.

18. (a) $(-7\pm\sqrt{21},-3)$, (b) $(-7\pm\sqrt{15},-3)$, (c) $x=-7$, $y=-3$.

Chapter 10, Exercise 5

1. F: $(3,2)$, $V(1,2)$, $\mathcal{D}$: $x=-1$.

2. F: $(-1,-6)$, $V(-1,-2)$, $\mathcal{D}$: $y=2$.

3. F: $\left(-7\dfrac{1}{80},-1\right)$, $V(-7,-1)$, $\mathcal{D}$: $x=-6\dfrac{79}{80}$.

4. F: $(90, -74.5)$, $V(90, -75)$, $\mathcal{D}$: $y = -75.5$.

5. $e = \frac{1}{2}$, F: $(\pm 3, 2)$, $\mathcal{D}$: $x = \pm 12$.

6. $e = \frac{3}{5}$, F: $(-4, 2)$, $(-4, 8)$, $\mathcal{D}$: $y = \frac{40}{3}$, $y = -\frac{10}{3}$.

7. $e = 3$, F: $(1, -1)$, $(-5, -1)$, $\mathcal{D}$: $x = -\frac{5}{3}$, $x = -\frac{7}{3}$.

8. $e = \frac{5}{4}$, F: $(3, 9)$, $(3, -1)$, $\mathcal{D}$: $y = \frac{36}{5}$, $y = \frac{4}{5}$.

9. $e = \sqrt{3}$, F: $(-2, 3 \pm 2\sqrt{3})$, $\mathcal{D}$: $y = 3 \pm 2\frac{\sqrt{3}}{3}$.

10. $e = \frac{1}{2}$, F: $(1, -3)$, $(1, 1)$, $\mathcal{D}$: $y = 7$, $y = -9$.

12. $x^2 - 4y^2 = 80$.

13. $24y^2 - x^2 = 2400$.

14. $5x^2 - 4y^2 = 20$.

15. $5x^2 + 9y^2 = 180$.

16. $8x^2 + 3y^2 = 35$.

17. $x^2 - 4y^2 = 4$.

18. $5x^2 + 9y^2 = 45$.

19. $80x^2 + 81y^2 = 6480$.

Chapter 10, Exercise 6

1. $(1, 2)$, $(5, 2)$.

2. $(6, 4)$, $(-1, 4)$.

3. $(1, 10)$, $(4, 1)$.

4. $(-2, 5)$, $(3, 0)$.

5. $(0, 1)$, $(-2, 5)$.

6. $(-4, 34)$, $(5, 25)$.

7. $(3, 2)$.

8. $(1, 2)$.

9. $(2/3, 2)$.

10. $(-2, 4)$, $(1, 1)$, $(2, 0)$.

11. $(3, 7)$, The curve and the line are tangent at $(0, 4)$.

12. $(-2, -1)$, $(-1, 2)$, $(4, 17)$.

13. $(-1, 1)$, $(3, 3)$.

14. $(-3, 1)$, $(1, 3)$.

15. Four points $(\pm 3, \pm 2)$.

16. $(\pm 2\sqrt{3}, \pm 2)$.

17. $(1, 0)$, $(0, 1)$.

18. No solution.

19. $(-3/4, 29/2)$, $(1, 11)$, $(3, 7)$.

20. $r = 2$.

Index

Contents for Other Volumes

Volume 1

Volume 3

Volume 4

Volume 5